Mapping It Out

Chicago Guides
to *Writing*, Editing∧
and Publishing

Mapping It Out

Expository Cartography
for the
Humanities
and
Social Sciences

Mark Monmonier

The University of Chicago Press
Chicago & London

MARK MONMONIER is professor of geography in the Maxwell
School of Citizenship and Public Affairs at Syracuse University.

The University of Chicago Press, Chicago 60637
The University of Chicago Press, Ltd., London
©1993 by The University of Chicago
All rights reserved. Published 1993
Printed in the United States of America

02 01 00 99 98 97 96   2 3 4 5

ISBN: 0-226-53416-2 (cloth)
0-226-53417-0 (paper)

Library of Congress Cataloging-in-Publication Data

Monmonier, Mark
   Mapping it out: expository cartography for the humanities and
social sciences / Mark Monmonier.
   p.m. —(Chicago guides to writing, editing, and
publishing)
   Includes bibliographical references and index.
   1. Cartography. 2. Humanities—Graphic methods. 3. Social
sciences—Graphic methods. I. Title. II. Series.
GA108.7.M66 1993
912—dc20   92-39894
                                      CIP

To

DONALD MEINIG

*with appreciation
for inspiration and
encouragement*

Contents

Preface

The goal of this book is to encourage scholars to use maps where maps are needed. Although written language allows authors to announce goals, discuss sources, explain research strategies, narrate events, and summarize arguments, prose has a sequential, linear structure that can be painfully insufficient for discussing places, regions, and spatial relationships. Simply put, efficient, lucid communication often requires organizing the information in two dimensions and mapping it out. My hunch is that many (perhaps most) scholars already recognize the value of the map as an expository tool but need to be shown (or coached in) how to design and make one. Encouragement thus takes the form of demystifying map making by showing how straightforward graphic logic can help the scholar-author compose visually effective maps to improve the clarity and impact of a book or article.

My subtitle, *Expository Cartography for the Humanities and Social Sciences*, not only recognizes the similarity of expository writing and narrating with maps, but also signifies the book's focus on the less quantitative, more literary side of scholarly discourse. A geologist, ecologist, geophysicist, meteorologist, botanist, or paleontologist should find *Mapping It Out* useful, but each probably has cartographic needs and ambitions different from those of a humanist or social scientist. For example, the so-called hard disciplines are well ahead of the other (softer? gentler? kinder?) fields of scholarship in using contours or isolines to describe terrain, atmospheric pressure, or a wide variety of highly abstract interpolated statistical surfaces. Serving heterogeneous audiences is difficult at best, and I have chosen to focus on an array of disciplines with comparatively similar needs. Although journalists and creative writers may also benefit from *Mapping It Out*, I go well beyond the reporter's or the novelist's typical cartographic need for comparatively straightforward locator maps. Yet "precision journalists," creative authors of quality nonfiction, and other writers about politics, society, institutions, commerce, mass communications, places, travel, history, and the arts clearly could benefit from learning about and using expository cartography.

I don't mean to disparage journalists, whose frequent cartographic efforts and occasional triumphs remind any well-informed person of the map's ability to describe, narrate, and explain. In the mid-1980s, with the support of a Guggenheim fellowship, I took a year's leave to look at news maps, study their content and use, and talk to their creators and publishers. What I saw was almost always a group effort in which the reporter or feature writer was seldom the author of the map that accompanied an article, however closely he or she might have worked with the illustrator or graphics editor. New technology has vastly increased the use of news maps; but its immediate users have usually been artists, sometimes journalists, but only rarely reporters. The news business is driven by deadlines more pressing, obtrusive, and—well—deadly than those affecting scholarly writing, and these deadlines, as well as tradition, mandate specialization. I concluded my book *Maps with the News* by asserting that journalistic cartography's next revolution will have reporters playing an integral role in initiating and crafting maps and other information graphics. Yet this revolution in map authorship seems likely to reach scholars before journalists, because of the individuality of scholarship and the scholar's relative freedom from onerous deadlines.

"Map authorship," which is what this book is about, is more than a glib phrase. Maps used in scholarly books and articles should not only support and supplement the author's words, but should also convey a definite message. A good expository map is an integral and unambiguous part of the author's narrative, adapted to the reader's expectations and prior knowledge and designed to promote understanding, complement other illustrations, and build on what has gone before. Like good prose, an effective expository map requires careful thought and conscientious editing and rethinking. Because the expository map is a part of the author's narrative, it is best designed while the author is making notes, writing, and rewriting. Indeed, writing with maps and with words should be a holistic process in which the author uses words and graphic symbols in concert and in combinations appropriate to the geographic character of the phenomenon discussed.

Mapping It Out is neither a textbook nor a critique nor a celebration of the map as a mode of communication. A textbook would have to be more rigorously comprehensive and more like a manual—and less fun for me to write and for you to read. Cartographic criticism can be useful at times, but promoting map authorship by ridiculing works having no maps or poor maps would be self-defeating. Scarcely more useful would be a catalog of cartographic exemplars profusely praised and expensively reproduced—yet requiring skills or techniques beyond the grasp of the average scholar-author. Instead, my approach is to describe and demonstrate basic principles of authoring maps, great or mundane. But although most of my illustrations focus largely on specific principles of graphic logic, I could not resist including a few examples—some good, some mediocre but nonetheless effective—of ways prominent scholars have exploited maps to the benefit of both their readers and their discipline.

The book moves from basic principles to specific cartographic contexts. Chapter 1 examines briefly how the map provides humanists and social scientists with a spatial language for describing locations, discussing places, and interpreting two-dimensional arrangements of features. Chapter 2 discusses scale, map projection, cartographic generalization, and the need to tailor geometric distortion and feature selection to the map's goals. Chapter 3 presents a graphic logic for matching map symbols to data and communication goals and examines designs appropriate for six typical cartographic goals in scholarly writing. Chapter 4 looks at typography and the role of text in map titles, keys, place names, and symbols. Chapter 5 addresses a range of topics concerned with map compilation, including cartobibliographies, copyright and permissions, facsimile reproduction, and the evaluation of source materials. Chapter 6 looks at the design of maps portraying quantitative data. Chapter 7 explores various designs for mapping movement and change. Chapter 8 shows how scholars can use maps to detect and describe spatial patterns, portray association and correlation, and enhance the integration of diagrams, photographs, and verbal description.

Don't expect instructions for either graphic software use or traditional pen-and-ink cartographic drafting. That level of specificity would muddy my message and truncate this book's useful life. Knowing how to draw with an ink pen or an electronic mouse is, after all, no more essential to creating a map than knowing how to set type or to make printing plates is a prerequisite for writing a book. Moreover, electronic graphic technology is evolving rapidly, and software developers regularly offer upgraded versions, so that the useful half-life of how-to manuals currently has been reduced to a bit less than two years. Some computer-literate readers will already know how to use appropriate software; up-to-date guidebooks and improved graphic user interfaces can help others master the simple drawing functions used to create most of the maps in this book. Although electronic drafting might even now account for nine-tenths of all maps in scholarly publications, I would not discourage use of traditional methods by those with a suitable combination of persistence and talent. Yet in deference to the reader's legitimate concerns with "how-to-go-about-it," I include three appendixes to compare electronic graphics with traditional technology, discuss working with a university cartographic laboratory or a freelance cartographer, and describe briefly a number of useful books on information graphics, map design, and traditional pen-and-ink map making.

Some college instructors of cartography may use this book as a text. That's fine: it spreads my message about the importance of mapping as a part of writing. But this book's possible appeal as a supplementary text might lead to unfair comparisons with conventional texts, because of its focus, brevity, lack of color illustrations, and unabashed use of the pronoun "I." So let me reiterate: I wrote *Mapping It Out* as an essay to convince humanists, social scientists, journalists, and others who write about people, cultures, neighborhoods, or regions that there are times when their writing needs maps, and to show them how to begin. A full-scale text would not only miss my target audience but would duplicate much of what others already say quite well. My hope is that having read *Mapping It Out*, readers with somewhat complicated mapping needs will either peruse the textbooks listed in my bibliography (appendix C) or consult a cartographic illustrator (as discussed in appendix B).

I would like to thank Bill Strong for constructive advice on map copyrights, Syracuse University cartographers Mike Kirchoff and Marcia Harrington for assistance with the illustrations, and Marge and Jo for tolerance and encouragement.

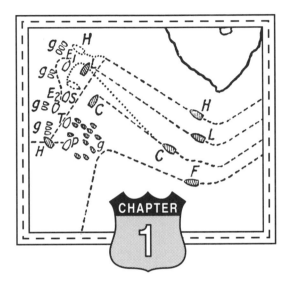

Maps in the
Humanities and
Social Sciences

MAPS HAVE AN IMPORTANT PLACE IN SCHOLARLY writing. Historians, sociologists, and other humanists and social scientists often write about territories and neighborhoods, about global disputes and local conflicts, and about causes and correlations involving areal differences, regional clusters, and other spatial patterns. By helping readers visualize regions and comprehend relative distances and other geographic relationships, maps amplify an author's sentences and paragraphs. After all, a two-dimensional stage may be more efficient than a one-dimensional trail of words for recreating and explaining a two-dimensional event. And symbols spread across a map can more effectively communicate the details and structure of neighborhoods, landscapes, and battle zones than words alone. Military strategists and urban planners need maps, and so do scholars whose subjects have any geographic aspect.

But not everyone who should use maps does. I am perpetually perplexed by the work of geographers who seem to have little interest in employing maps to communicate, interpret, or explain. At times, I ask myself if these colleagues are really practicing geography. And often it is clear that they aren't. Yet, as the holder of a Ph.D. in geography and a faculty position in a department of geography, I am often both pleased and puzzled that many noncredentialed geographers are doing interesting geography in other academic disciplines. Scholarship is not bound by the labels we use to mark territory at universities; nongeographers should practice geography if they choose, just as geographers should feel free to contribute to other social sciences, philosophy, the humanities, computer science, or statistics. Indeed, the domain of geographic scholarship is not only too broad for the meager community of researchers trained in the discipline, but also too important to be limited to people with geography degrees.

The misuses of maps amaze and even delight me. The map is a robust medium, and even bad maps may communicate, albeit crudely and inefficiently. That noncredentialed geographic scholars may seem compelled to use them attests to the map's inherent

3

role in "earth writing"—the literal meaning of "geography." But it is astonishing that careful writers who have spent considerable time planning, sculpturing, and polishing their prose often have little appreciation that mapping, like writing, can be done lucidly and elegantly. Helping the conscientious scholar create and use visually efficient, aesthetically satisfactory maps is my goal in this book.

This chapter begins with a brief commentary on the limitations of verbal discussion and on the neglect of cartographic illustration in master's and doctoral programs in the humanities and social sciences. It then examines location and spatial pattern as elements in scholarly work, and the consequent need for maps.

WORDS AND MAPS

As *National Geographic* has demonstrated for decades, maps and other pictures help explorers share with readers their insights and discoveries about both large and minute parts of the world. Humanists and social scientists are explorers, too, and many are geographers in spirit if not in disciplinary affiliation. Because their explorations touch several aspects of place and space, maps can have an important role in their writing. For instance, the literary scholar focusing on Dickens needs to develop and share a broader, more concrete knowledge of the scale and structure of nineteenth-century London than Dickens's classic novels provide. Similarly, the medieval historian might need to know and communicate not only the locations of fortresses and monasteries, but also the theologically influenced cosmological-cartographic world view of twelfth-century nobility. And the student of Napoleon needs to appreciate and explain the effects of the terrain and climate of the Russian steppes, as well as the tribal diversity and economic resources of eighteenth- and early nineteenth-century Europe. The list of non-credentialed geographers is long and includes the anthropologist, the archaeologist, the art historian, the economist, the literary scholar, the political scientist, and the sociologist.

Like a writer for *National Geographic*, an academic explorer needs to appreciate the marvelous capacity of the eye-brain system for processing pictorial, two-dimensional data. Photographs and other pictorial illustrations allow the reader to see what the explorer saw, at least from an insightfully selected vantage point.

The reader can form mental images that foster comprehension and understanding. Scholarly pursuits often call for more complex pictures, such as the spatially meaningful arrangements of abstract symbols on maps and statistical diagrams. These images help the reader see how the academic explorer has organized, processed, analyzed, or interpreted observations and measurements. The author who relies solely upon words may hobble the reader by obscuring facts and hiding information.

Consider two examples from the work of military historian, naval officer, and sea-power advocate Alfred T. Mahan, whose writings include the two-volume *Sea Power and Its Relationship to the War of 1812*. Mahan used numerous maps to explain the geographic settings and the choreography of naval engagements. Figure 1.1, which reconstructs the costly victory of the American warship *Constitution* over the British frigate *Java* off the coast of Brazil on December 29, 1812, illustrates how maps complement verbal description. Mahan tells his readers that this was not a typical artillery duel but "a succession of evolutions resembling the changes of position, the retreats and advances, of a fencing or boxing match, in which the opponents work round the ring." His map both dramatizes and documents this spatial complexity, as each ship is shown maneuvering to attack with more of its guns facing fewer of its opponent's guns. An arrow indicates wind direction, contrasting boatlike symbols differentiate the two ships, dashed lines portray their courses, and numbers show the time in hours and minutes for various simultaneous positions between

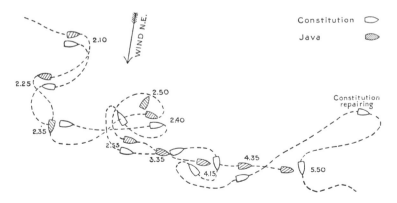

FIGURE 1.1. "Plan of the engagement between Constitution and Java."

2:10 p.m. and 5:50 p.m. Although the *Java* damaged the American ship's wheel, the *Constitution*'s guns destroyed all but one of her opponent's masts. After moving away around 4:35 to repair its own damage, the *Constitution* returned at 5:50, and the British surrendered. By helping the reader organize two pages of details about individual attacks and their effects, the map makes its author's words more comprehensible and convincing.

Not all of Mahan's illustrations address events on a featureless sea. Figure 1.2, a less dramatic but more graphically complex

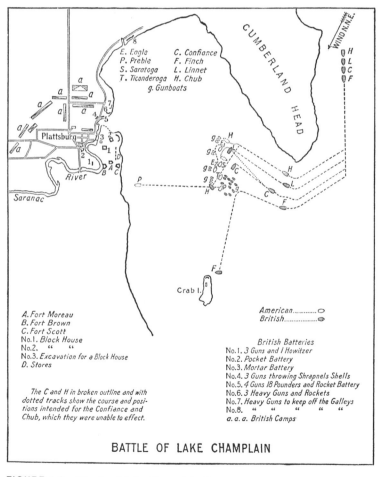

FIGURE 1.2. "Battle of Lake Champlain."

map, provides the geographic framework for a 24-page examination of the "decisive" Battle of Lake Champlain on September 11,
1814. Mahan's text describes the weeks preceding the battle, as
British troops under Sir George Prevost advanced southward
along the western shore of the lake, sacked Plattsburg, sur-
rounded the town, and set up heavy guns on the Saranac River
opposite the group of American forts and blockhouses identified
on the map. Contrasting open and shaded boat symbols differentiate the American and British ships, smaller open boat symbols
indicate American gunboats, and letters and numbers identify
individual warships and important land positions. Solid lines
show the shore of Lake Champlain, and dashed lines portray the
paths of individual ships. Both sides had prepared a strategy,
with each vessel assigned to engage a particular opponent, and
dotted lines represent the intended courses of three of the four
British ships. The American commander Thomas Macdonough
positioned his vessels at the entrance to Plattsburg Bay, out of
the range of British cannon and sufficiently far north that the
British had to approach head on. An arrow indicates the north-
northeast wind that brought the British ships around the tip of
Cumberland Head into the bay but failed to carry them to their
assigned positions. By maneuvering the *Eagle* and the *Saratoga*
so that most of their guns were on the "engaged" side, the
Americans gained an advantage over the British, who had also
counted on support from their heavy artillery at Plattsburg. By
the end of this intense two-and-a-half-hour battle, the *Finch* had
retreated to Crab Island, and the other three British vessels had
surrendered. Mahan's use of this and other maps reflects not
only a military commander's experience with battle plans but
also a historian's concern with organized and illuminating communication of important details.

Integrating maps and words is the focus of this essay, not the
question of whether one is better than the other. When an article,
book, or dissertation concerns interaction among places, words
with maps can be far more powerful as a vehicle for scholarly
exposition than the same words *without* maps. Scholarly writing
always has and always will depend largely on words. To be useful, maps must complement our sentences and paragraphs, not
compete with them. The committed scholar must learn when and
how to use all relevant communication tools, graphic as well as
literary, and how to employ them in concert, not in isolation.

Most academic researchers—even some geographers, sad to say—know little about making maps. Most undergraduate courses fail to address, much less advocate, the possibilities of employing graphics to explain spatial concepts, and anthropologists, historians, and others who could make frequent and effective use of maps rarely study cartography. Graduate programs in the humanities and social sciences ignore map making as an analytical and expository skill. Indeed, graduate training and disciplinary tradition have treated map making as a service that one buys, rather than as a potentially important part of the scholar's creative work. Thus, when maps are used at all, it seems, someone else draws them, perhaps an illustrator hired by the university press but more likely a free-lancing undergraduate majoring in geography, art, or architecture. The "map author" often relies too heavily on the cartographic advice of an assistant whose only asset is superior hand-eye coordination. The resulting maps at best are visually pleasant, decorative props rather than important supporting players. And at worst, they distract, confuse, or mislead the reader.

Neglect of expository communication in master's and doctoral training programs partly explains the neglect of mapping in the humanities and social sciences. Academic disciplines live and grow by collecting and organizing facts and by developing, refining, and sometimes purging theories to explain these facts. Training programs quite naturally emphasize observing, sampling, and theorizing, and sometimes also a particular research skill such as statistical analysis, computer programming, or reading German. A single course on research methodology might address bibliographic sources, data-collection strategies, and proposal writing. Except in creative-writing programs, graduate faculty presume (often naively) that their students have learned to write as undergraduates. Tradition and a focus on current paradigms preempt formal instruction in writing and other communication skills; professorial mentors rely largely on occasional marginal reminders of the value of topic sentences, active voice, and concise wording. Good writing habits may be reinforced by such informal admonishments, but students usually have no previous academic experience in cartography to rely upon or recall.

Western society's word-oriented view of literacy and expository expression also partly accounts for the limited use of maps by humanists and social scientists. Although reading and writing are no doubt the paramount communication skills of the educated person, the complete scholar should cultivate three important companions to literacy: articulacy, numeracy, and graphicacy.

Articulacy refers to fluency in oral expression and relates to both stage presence and command of language. Sometimes training in articulacy is a formal undergraduate requirement, under the course title "Public Speaking." It may be particularly prized by scholars who want to share their knowledge and enthusiasm with public-television audiences or large-enrollment lecture classes. Other scholars often regard a highly articulate colleague with a mixture of envy and suspicion, considering him or her gifted but perhaps too entertaining, and thus somewhat shallow.

Numeracy refers to fluency in the manipulation of numbers. American educators have a supportive but restrained view of numeracy: school boards are pleased if high-school graduates can balance a checkbook, and college faculties insist that students at least have a minimal exposure to algebra and trigonometry. Among the humanities and social sciences, respect for numerical competence varies widely with the discipline's ability to generate or exploit quantitative data. In the postwar era, "quantitative revolutions" occasioned much paradigmatic blood-letting in disciplines in which "quantifiers" and "nonquantifiers" fought for power in the name of principle. Even today, some humanists condemn as too narrowly reductionist any research employing counts and averages, and some social scientists steeped in highly abstract mathematics regard as trivially empiricist studies that are based on actual numbers.

Graphicacy refers to fluency with graphs, maps, diagrams, and photographs. The most spatial of the four groups of communication skills, graphicacy has asserted itself in schools and colleges in courses in commercial art, illustration, and mechanical drawing, all of which apparently are too technical and not relevant enough for the needs of liberal arts majors. Yet even some inherently graphic disciplines ignore the value of graphics to organize and elucidate: statistics majors often emerge with no training in exploratory graphical data analysis, for instance, and some geography majors are not offered or required to take a course in cartography.

Attempts to increase the amount of graphic material in publications can provoke bitter debate. For instance, American newspaper publishers who redesigned their papers in the 1970s to counter circulation losses blamed on television opened an old and sometimes heated journalistic conflict between "word people," who couldn't care less about news photos and other editorial artwork, and "picture people," who had a more catholic view of news presentation. Since there were limited budgets for both personnel and space in the publication, what one person or group gained could be someone else's loss; competition for space and influence often reflected competition for salaries and staff. Thus, a decision to use more information graphics commonly increased the number of artists and the salary and prestige of the art director, possibly at the expense of reporters and junior news editors. The attempts that were made to explain such struggles with intriguing left-brain/right-brain hypotheses ignored the tenuous origin of brain-hemisphere theories in clinical studies of brain-damaged accident victims. In reality, the journalistic furor was a minor debate between stubborn traditionalists and those less conservative in the adoption of new technology. Publishers, owners, and senior editors decided in favor of more graphics because surveys indicated that consumers wanted a more visually appealing, better organized, more readily understood newspaper. Although *USA Today* received much attention for its spectacular color printing and editorial artwork, the *New York Times, Christian Science Monitor, Washington Post*, and other "elite" newspapers had increased their use of information graphics even before the appearance of their more colorful competitor.

Technology clearly was the driving force. Before computers, electronic publishing, and graphics networks, the print media used fewer information graphics largely because they cost too much and took too long to produce. Newspaper firms are businesses, after all, and news is a perishable commodity. That newspapers at one time used woodblock engraving and other tedious techniques to illustrate with maps their accounts of important battles attests to the expository power of two-dimensional representations. That newspapers used comparatively little editorial art attests to journalism's word-oriented origins in the "news-letters" of the fifteenth and sixteenth centuries. Advances in photographic engraving in the 1880s, electronic facsimile transmission in the 1930s, and computer graphics in the 1980s allowed news publishers to add more graphics and to more effectively explain spatial events in a spatial format.

Despite successes in integrating words and graphics in newspapers and weekly news magazines, the print media have yet to integrate words and graphics in the creative routine of working journalists. Ever since the cottage-industry newspaper became a mass-communications enterprise, news publishing has relied upon a highly structured, compartmentalized labor force. Indeed, one of the newspaper executive's more demanding responsibilities has long been to negotiate with and referee jurisdictional disputes among more than a dozen specialized craft unions. Although electronic technology has replaced the typesetter and given editors fuller control over the work of reporters and the layout of the paper, writers still almost always do the writing while artists do the drawing. To be sure, quality control is no longer an obstacle, for graphics software and a rich library of "clip art" provide aesthetic support and the editor and the art director can easily sharpen and reformat the reporter's self-composed graphics. This technical and professional support for grass-roots explanatory graphics should yield more accurate reporting and more lucid writing.

Some large newspapers have recognized the need for increased integration of words and graphics by creating a new specialist, the graphics editor or graphics coordinator. An experienced journalist who acts as liaison and broker between reporters and artists, the graphics coordinator tries to anticipate by at least several hours the need for maps and diagrams and to make certain that the artwork agrees with and reinforces the story. In this scheme, the reporter might collect and annotate maps, make pencil sketches, and even collaborate face-to-face with the artist, but someone else designs and draws the map, even a very simple one. The graphics revolution in news publishing needs two institutional breakthroughs to finally win the battle of representational integration: each reporter needs to be provided with a competent graphics workstation; and a broader view needs to develop of the journalist's responsibilities and necessary skills.

Despite a slower start, humanists and social scientists are freer than media journalists to fully integrate words and graphics in their thinking and writing. The scholar is not burdened by the institutional baggage of a complex business divided into specialized departments linked to a single product, but can be seen as an entrepreneur running a one-person small business of sorts. Taking full advantage of powerful yet inexpensive electronic publishing systems is therefore easier. The academic researcher can create and manufacture a manuscript by processing words, graphics, bibliographic information, and numerical data on a personal computer. Even on big projects involving several collaborators or assistants, the separation of words and graphics need be no more pronounced than the division of labor for analyzing data and for writing and editing reports.

Yet graphics software no more guarantees good maps than word-processing software assures good writing. Artistic ability does not imply cartographic skill, and making maps well, even simple maps, requires training or at least some informal study. Like expository writing, cartography is not wholly intuitive. Cartographic symbols have a unique vocabulary, logical rules that promote efficient, unambiguous decoding, and stylistic conventions that reflect both pragmatism and aesthetic biases. Cartographic grammar might not be as well developed as linguistic grammar, but some combinations of data and symbols work better than others, some combinations don't work at all, and some combinations can easily mislead the ignorant map author, as well as the naive map reader.

Beware of software products that promise instant maps. Unfortunately for many would-be map makers, not all developers of mapping software are aware of the principles of cartographic design. And unfortunately for many software users, it is possible to produce an attractive, well-balanced map with neat symbols and crisp labels that is a confusing, graphically illogical puzzle, useful perhaps for decoration but for very little else. In later chapters I discuss the basic concepts of cartographic representation and present design strategies useful to writers who want to explain their analyses, interpretations, and ideas with simple maps. Like those of most grammars, the cartographic rules are usually straightforward and logical, yet often unforgiving when ignored.

Maps, Location, and Spatial Pattern

Writing with maps works best if the scholar learns to think spatially and to use maps at all stages of research, not just while writing. Mapping, after all, is not solely a medium for communication, but is also a tool of analysis and discovery. So if we want to optimize our use of maps and to gain whatever insights they may hold for us or our readers, we need consciously to search for maps while researching in the archives or conducting interviews, or to annotate them while observing subjects' behavior or studying our data. Maps work best for organizing information if we condition ourselves to look for information worth mapping.

Among the many social scientists and humanists who have found the map a useful research tool is Robert Park, a sociologist by disciplinary affiliation but clearly a geographer by instinct. Among Park's more intriguing publications is *Old World Traits Transplanted*, cowritten with Herbert A. Miller and first published in 1921. This study explores the customs and institutions that European and Asian immigrants brought to American cities. Some of its maps describe and explain regional patterns. Figure 1.3, for example, uses open dots representing Roman Catholic churches and schools using the French language to show concentrations of French Canadian immigrants in the mill towns of New England and in areas immediately adjacent to Quebec. Despite its crudely drawn state boundaries, the map presents a instructive view of the French Canadian immigrant's preference for rural or small-city destinations close to Quebec. The absence of dots in Boston and New York City is as revealing as the clusters in north-

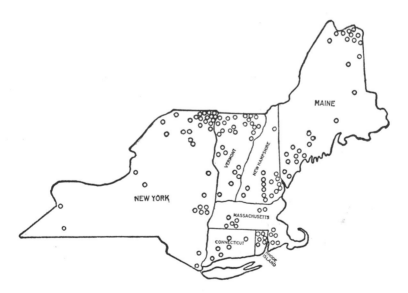

FIGURE 1.3. "French Canadian parishes of New England and New York."

ern Maine, northeastern New York, the Connecticut and Merri-
mack valleys, and the Narragansett Bay area.

Park and Miller included more city and neighborhood maps
than regional maps. Some of these maps portrayed concentrations, but others focused on diversity. Figure 1.4, for instance,
shows the wide variety of national origins of the founders of Jewish synagogues in a twenty-block area of lower Manhattan. Point
symbols differing in shape and darkness represent eleven different countries as well as an "unknown" category. The map is an
effective testament to the ethnic and linguistic complexity of the
early twentieth-century New York Jewish "community," in
which Levantine Jews were isolated by language from Yiddish-
speaking Jews. As the authors observed, "Only the Jews themselves appreciate how profound are these differences. While their
spiritual life is based on the same historic traditions, the different
groups have lived in different ghettos as separate, self-governing
communities, suspicious of any intrusion whatever into their affairs." But making a statement about the astonishing level of diversity is far less effective than demonstrating it with a map. Park
and Miller used other detailed maps to show distinct geographic
differences among Manhattan's various "Italian colonies" in
their residents' places of origin within Italy.

As these examples suggest, location and spatial pattern are
important elements in the description, explanation, or interpretation of many phenomena that interest scholars. Moreover, although we might prefer not to admit it, at least part of what most
social scientists and humanists do is journalism, albeit with a
thoroughness, thoughtfulness, and attention to detail that the
profit-motivated, deadline-driven news media would not tolerate. The similarity to journalism lies in six basic questions scholars ask: Who? What? When? Where? Why or how? And, so
what? Of course, geographers might pay more attention to
"where," and historians to "when," and sociologists and anthropologists to "who," but from time to time all six questions arise
for all of us, with the first four providing a foundation for addressing the last two.

The journalist's first four questions arise early in large projects
requiring an electronic database. The researcher must decide
what entities, attributes, and relationships to include in designing
a database. These database concepts—entity, attribute, and relationship—are not meaningless jargon; all areas of scholarship,

especially the humanities, are rapidly recognizing their importance. Briefly, an *entity* is an object or event described by one or more *attributes* and linked to other entities by one or more *relationships*. "Where" might arise in any of the database elements. Some entities, for instance, might be cities, with attributes such as population size, land area, type of government, and date of incor-

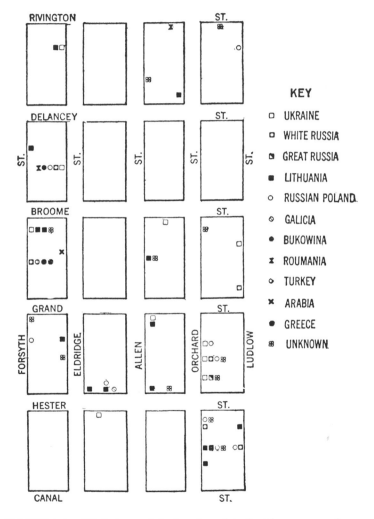

FIGURE 1.4. "Birthplaces of the founders of the Jewish synagogues in a congested New York district."

poration. A "within" relationship might link the city to one or more states, and an "including" relationship might point to the census tracts that subdivide the city. These three kinds of entity (state, city, census tract) might have as locational attributes the latitude and longitude of a single center point or a list of coordinates that describes the length and shape of the entity's boundary. At a finer level of detail, the schedule of entities might include individual buildings—say, an important writer's birthplace, elementary school, place of first employment, and favorite bar. Each building's street-address attribute allows a "street-distance" operator to link these structures to an electronic representation of the street network, so that the system can then define a "minimum distance" relationship linking pairs of such entities. Although many of us will be spared the complex task of designing an electronic database, all researchers should determine whether spatial entities, locational attributes, or geographic relationships might be important, and if so, devise a strategy for including them.

Most readers of this book probably will not need to develop their own highly structured electronic database. Yet recording addresses, place names, geographic coordinates, and other locational identifiers while collecting other data can be valuable if there is the slightest chance that there will be a need to map the data. The goal, of course, should not be merely to know where individual events occurred, but to recognize possibly meaningful geographic patterns—for example, a similarity between annual rainfall and population density, or a decline in land value with increased distance outward from a city center. Although such relationships are hardly unexpected and might even seem trivial, the map might also reveal areas where general, logical trends do not hold. Such anomalies—potentially meaningful exceptions to an expected pattern—can suggest additional, heretofore unknown causal factors. Epidemiologists have long recognized the value of mapping data, and the National Institutes of Health devotes a small but important part of its budget to mapping death rates for various kinds of cancer in a search for hot spots possibly related to the presence of high levels of radiation, hazardous materials, heavy-metal contamination of drinking water or air, or other environmental factors. Good research requires thoroughness.

Maps can also point out what we don't yet know and still need to look for. In conducting a survey of local governments, for instance, a map might be used as a checklist to show which municipalities did not respond to an initial questionnaire. The map might reveal a pattern of nonresponse related to distance, type of government, or party affiliation. While suggesting the need for a carefully focused follow-up, the map might also be useful for planning whatever face-to-face approaches are required to complete the survey. Plotting critical responses on maps can suggest the need to enlarge the study area or indicate the efficiency of intensive sampling in only a few key places. A map can summarize what is known and not known about a city, neighborhood, or region; the blank spaces and question marks that emerge after recording what the existing literature tells us can suggest where further inquiries might productively be focused. Geologists often use maps to determine where to sample next, and sociologists, anthropologists, historians, and others can benefit from a similar strategy. Also, composite maps of what the discipline as a whole knows can reveal inconsistencies demanding detailed study. I have argued that geographers who want to work on important questions might begin by composing the table of contents of a good thematic atlas and then noting those sections or individual maps for which reliable information is lacking. As a form of exploration, geographic research needs to probe a variety of frontiers, some continental or regional, but some in our cities, and many in our neighborhoods.

Appropriate strategies for collecting and analyzing data vary widely among the readers I hope to reach with this book. Economists, sociologists, and others who work with censuses and surveys should look carefully at their data before calculating rates or computing statistical correlations. For data already identified by location or aggregated to states or census tracts, mapping is an essential part of *exploratory data analysis*, a collection of statistical and graphic techniques particularly useful in confronting poorly structured problems in an information-rich environment. Recognition of distance relationships, density variations, and similarities with other variables can be a productive path to a higher level of understanding.

Humanists and social scientists who collect their own data may find the map a convenient framework for taking notes or organizing field observations. Research in the humanities is often a holistic endeavor, and if location is at all a factor, the map provides a useful structure for allowing isolated facts to form a pattern, either among themselves or in relation to other geographic features. In addition to supporting a systematic cataloging of geographic facts and hunches, the map can stimulate serendipitous discovery.

Gathering data might also include collecting maps. In addition to providing information about a past geography, maps in archives can prompt useful insights about map makers and the cultures and societies in which they worked. For example, older published or manuscript maps sometimes suggest to the critical scholar how governments, explorers, writers, or earlier researchers saw an area themselves, or how they chose to represent it to others. Humanists might want to share with their readers facsimiles of representative or important maps, or to use cartographic artifacts as evidence of propaganda or biased interpretations.

For all of the reasons mentioned in this section for using maps to collect, organize, and analyze information and to communicate spatial facts and relationships to readers, awareness of maps while carrying out the research and preparing the manuscript will improve an author's ability to recognize, understand, and discuss geographic concepts and details. Although maps added after writing the first draft can be beneficial to both author and reader, earlier awareness of cartographic information and cartographic concepts can assure a fuller, more coherent integration of maps with words. By focusing the following essays on principles and practices particularly relevant to the humanist or social scientist, I hope to elevate the typical reader's limited awareness to a working knowledge of the full role of maps in scholarly communication.

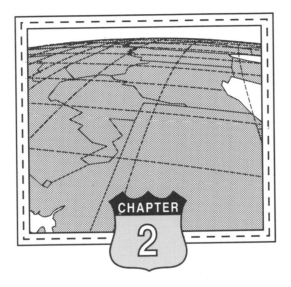

CHAPTER

2

Scale, Perspective, and Generalization

MAPS ARE SCALE MODELS OF REALITY. THAT IS, THE
map almost always is smaller than the space it represents. I say
"almost" because the graphic representations of subatomic par-
ticles and chromosomes drawn by nuclear physicists and geneti-
cists are indeed maps, in the broad sense of using visible marks to
portray relative positions. At the other end of the spectrum of
map scales are the astronomer's star charts, which represent the
space-time relationships of an expanding universe. Like the mi-
croscope and the telescope, the geographic map can be an instru-
ment of observation and discovery. And like genetic maps and
galactic maps, geographic maps model the sizes, distances, and
relative locations of phenomena we believe are real. By allowing
us to discover or impose structure, these cartographic models
promote both understanding and communication.

As a graphic interface between reality and the mind, the map
presents a selective view of reality—selective in the space it portrays, the viewpoint it offers, the objects it includes, and the symbols it uses to represent these objects. The map author must must
make choices in three main elements of this graphic interface:
scale, projection, and symbolization. *Scale* refers to the degree of
reduction and is commonly stated as a ratio of distance on the
map to distance on the ground. Because the scale of most maps is
in the range between 1:5,000 and 1:500,000,000, cartographic
models require considerably more generalization and abstrac-
tion than, say, a model railroad, with a scale between 1:48 and
1:220. *Projection* refers to the mathematical transformation that
assigns objects on a curved, three-dimensional surface to locations on a flat, two-dimensional plane. Map projection tends to
distort scale, distance, and area, and the projection chosen can
either serve or thwart the map author's goals. This chapter examines these first two cartographic elements, and the following
chapter addresses map symbols, a more complex subject with a
wider variety of choices.

Map projection can be a difficult subject, and readers who did
poorly in school geometry may find this chapter's second and

third sections a bit overwhelming at times. Projection is an essential element of all maps; although these sections offer an appropriately simplified treatment, there is no way to avoid altogether a discussion of how geometric distortion affects the map maker's choices. So skim-read if you must, but at least look carefully at the illustrations and their captions, and make certain you understand the guidelines at the end of the chapter. Subsequent chapters will repay your perseverance by returning to a comparatively straightforward, more obviously pragmatic discourse, like that of the first section's treatment of map scale.

Scale Models and the Representation of Scale

As scale models representing landscape features and other real phenomena with graphic symbols, all maps are generalizations. In most cases generalization results because the map cannot portray reality at a reduced scale without a loss of detail. Yet cartographic generalization can reinforce the communication goal of a map author who wants to highlight some features or details and deemphasize others. For example, a map author might filter out minor indentations in a coastline in order to call attention to its dominant trends. In other cases generalization might require deliberate exaggeration of important but easily overlooked parts of the coastline. On small-scale maps of New England, for example, exaggeration to point out Cape Cod and a few of Maine's drowned valleys might even be essential to make the map "look right."

On almost all maps, graphic symbols demand proportionately more space than would precisely scaled-down representations of the objects they portray. Consider, for example, the typical 1:24,000-scale topographic map showing roads, boundaries, elevation contours, buildings, and similar details about infrastructure, terrain, and political jurisdiction. At this scale, a double-dash line symbol 1/50-inch wide representing a one-lane unimproved dirt road about eight feet wide preempts a graphic corridor on the map that could exactly portray a feature on the ground forty feet wide. This exaggeration is necessary for visibility, because a true-scale symbol only 1/250-inch wide would be difficult to see and too thin to print. Exaggerated width is essential if cartographic line symbols are to use differences in color, width, pattern, and texture to distinguish among a variety of boundaries, transportation routes, and other linear features. Smaller scales require even greater symbolic exaggeration; at 1:2,000,000, for instance, a thin red double-line symbol about 1/60-inch wide representing an interstate highway consumes as much space as a corridor a half-mile wide. Even greater exaggeration occurs where visible lines portray the location and shape of boundaries, elevation contours, and other invisible features of infinitesimal width.

Significant features are often clustered together, so map generalization requires not merely selection and symbolic exaggeration but also simplification, smoothing, displacement, and classification. Thus, a topographic map might simplify an L-shaped house to a rectangle; smooth out the minor bends of a meandering stream or jagged shoreline; displace a road, railway, stream, and political boundary that are crammed together in a gap through a ridge; and assign each road to one of only five categories.

In general, the more compressed the scale of the map, the more severe the distortion required for a clear picture of important relationships. Figure 2.1 illustrates how map scale affects cartographic generalization. Portions of three topographic maps published at scales of 1:25,000, 1:100,000, and 1:250,000 all show roughly the same area around Henry David Thoreau's Walden Pond. At 1:25,000, the upper map offers the most detailed description of the pond's waterline and surroundings. Closely spaced contour lines encircling the pond indicate a moderately steep descent to the water's edge. The map identifies the steep area on the south as Emerson's Cliff. Technological intrusions on Thoreau's old neighborhood include a multitrack railroad to the southwest, a high-voltage power line to the southeast, and the Concord Turnpike to the north. At the east end of the pond three buildings stand between a paved road and a notch in the waterline. At 1:100,000, the center map still identifies the landmark by name but presents a simpler and smoother description of its shoreline. At 1:250,000, the lower map shows but does not identify the pond. As map scale decreases and the space representing an area shrinks, clarity demands the omission of less important features and their labels and a simplified geometry for those that remain. The need for careful selection of relevant features is but one reason I call the person who makes the decisions a map author.

In many cases, if not most, the map author's choice of publication and topic influence or dictate the scale of the map. Scale is a
two-component relationship between map size and territory size.
The overall design and page size of the intended publication usually limit the size of the map, whereas the author's topic deter-
mines the map's *geographic scope*, that is, the extent and size of
the territory the map must accommodate. For example, a historian writing for a university press might require a reference map of
modern France, which in the standard north-at-the-top orientation defines a square territory roughly 600 miles (970 km) on the
side. If the publisher prefers to have all artwork fit within the
book's standard *type page*, and if this rectangular area is 23 picas
(3.83 inches or 9.74 cm) wide by 39 picas (6.5 inches or 16.5 cm)

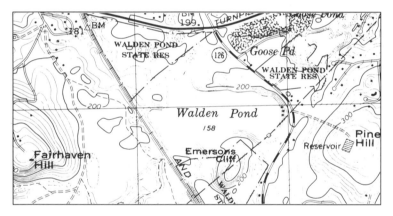

1:25,000

1:100,000

1:250,000

FIGURE 2.1. Walden Pond, as shown on modern maps at different scales.

deep, the horizontal dimension will limit the size of the illustration unless the map title, key, or caption is extraordinarily large. (Editors and printers often measure length in *picas*, of which there are exactly six to the inch.) If a rectangular border, or *neat line*, is to enclose the map, an additional quarter inch of space on each side would provide a graphic buffer between the neat line and the national boundary. The map must then compress France's 600-mile east–west extent to a published width of roughly 3.3 inches (8.4 cm), so the maximum scale is set at approximately 1:12,000,000. For a region elongated from east to west, such as Switzerland, Pennsylvania, or the forty-eight conterminous United States, turning the map sideways on the page would allow a slightly larger, somewhat more detailed map.

Scholarly journals usually have a somewhat wider type page. Typical widths for single-column type pages are 31 picas (5.17 inches or 13.1 cm) for the *Journal of American History* and 26 picas (4.33 inches or 11.0 cm) for *Technology and Culture*. In contrast, *Historical Archaeology* can accommodate a one-column map 17 picas (2.83 inches or 7.19 cm) wide or a two-column map 35 picas (5.83 inches or 14.82 cm) wide.

Before going further, I must define a few terms and explain in more detail the calculation of map scale. If 600 miles must fit within 3.3 inches, we can state a *verbal scale* of "one inch represents 180 miles," calculated by dividing the 600 miles by the 3.3 inches and rounding slightly. Some map users prefer verbal references relating familiar publication-size units to familiar world-size units, and the British often refer to their various national map series with such informal, approximate labels as "four-miles-to-the-inch" for the 1:250,000 series, also called the "quarter-inch" series because a quarter inch represents slightly less than one mile. Similarly, users of American maps can conveniently remember that on the U.S. Geological Survey's 1:24,000-scale topographic maps an inch represents exactly 2,000 feet. An older American series, phased out in the late 1950s, employs a scale of 1:62,500, on which one inch represents slightly less than one mile. Beware, though, of the colloquial sloppiness of some architects, planners, and city officials who omit all units and refer to 1:4,800-scale maps as "1 to 400," because one inch represents 400 *feet*. Although a detailed site plan or an archaeological survey plot might have a scale of 1:400 or larger, such maps are comparatively rare, and there is an enormous difference between the map scales 1:400 and 1:4,800 in detail, content, compilation cost, and physical size or number of map sheets.

To avoid such confusion, cartographers prefer to use dimensionless scales stated as ratios or fractions. Thus, 1:4,800 and 1/4,800 both mean that an inch on the map represents 4,800 inches on the ground, or that a foot on the map represents a distance of 4,800 feet. By convention, the numerator of the fraction is always one. For the page-width map of France, the ratio scale would be computed by first converting 600 miles to 38,016,000 inches by multiplying successively by the 5,280 feet in a mile and the 12 inches in a foot, and then dividing by the corresponding east–west width of the French border, 3.3 inches. The resulting scale is 1:11,520,000, rounded to 1:12,000,000 in order to provide a slightly more generous separation between the national boundary and the neat line.

A cartographer would definitely consider this a small-scale map, because its fractional equivalent, 1/12,000,000, is very small relative to, say, 1/4,800, a scale used for some highly detailed urban maps. When comparing map scales, cartographers always refer to the fractional scale, not the size of the denominator or the number to the right of the colon. So that just as a 1/8 slice of pie is smaller than a 1/4 slice of the same pie, 1/100,000 is a smaller scale than 1/50,000. Geographers generally regard maps with scales of 1/250,000 or smaller as small-scale maps and maps with scales of 1/24,000 or larger as large-scale maps. A small-scale map accommodates less detail than a large-scale map and tends to portray a larger territory at a smaller size.

Failure to consider the ultimate size of the published map might lead to an awkwardly sparse map needing more detail to fill wasted space. But far more common is a visually cluttered map crammed with relevant information that is communicated poorly, if at all. Occasionally a small amount of additional space for a map can be gained through strategies such as turning it sideways on the page or extending it either into the margin or across two pages. A "double-page spread" is visually awkward for the reader, because important details near the center of the map can be lost in the *gutter* between the pages. A highly persuasive author might convince the publisher to include a fold-out map printed on a wider sheet of paper. But because these tipped-in illustrations are generally pasted in by hand, they can add substantially to manufacturing costs. A large folded sheet map can go into a pocket in a book's inside back cover or into a journal's mailing envelope, but these alternatives are also costly. If an important article requires a particularly large map, or if the author can subsidize the added production expense, some scholarly journals might accept oversize material. But don't assume without prior negotiation that a book publisher will be willing to pass the added cost along to buyers through a higher cover price.

To adapt to a single-page format, the map author usually must eliminate some details, spread the information over several maps, or rely on a mixture of multiple maps and fewer details. A single-base, multiple-map strategy uses several maps at the same scale to show the distributions of different features through the entire study area. Some of these maps might contain only a single distribution, in addition to needed reference features, but combining two or more less dense or nonoverlapping distributions on a single map can save space, as can grouping together in a single drawing several smaller-scale maps of low-density information requiring less detail. An alternative is to partition regionally, perhaps by dividing the territory into either a grid of rectangular regions portrayed at the same scale, or a set of irregularly shaped subregions perhaps portrayed at different scales. Grid partitioning might not be suitable if important features or relationships extend across arbitrary quadrangle boundaries. In contrast, dividing the study area into more or less natural economic, cultural, or landscape regions can promote a more efficient use of space, because the density of important features frequently varies from region to region. Thus, a map with a somewhat larger scale might portray a small region having many densely packed features, while a single, smaller-scale drawing might accommodate two or more larger, yet less densely packed regions.

Partitioning the study area into regions that make sense to the reader also promotes the integration of maps and words. Indeed, in adapting to a single-page format by partitioning areally or among features, the map author should consider carefully the sequence in which the written text will discuss the supporting illustrations. In addition to the regional maps, an integrated presentation might also require one or more smaller-scale maps of the entire study region, as an introduction or a summary.

Even a map that fits neatly within a single page might have one or two areas requiring a *detail inset map* to portray at a somewhat larger scale a complicated area with dense symbols. For instance, a 1:12,000,000-scale, page-width map of France might well require a detail inset for Paris and its suburbs. Particularly dense concentrations are common in maps of a wide range of phenomena, such as military engagements, social movements, and industrial innovation. Areas not sufficiently dense to require a separate, larger-scale map might be shown by one or more detail insets on a single map providing an overview for the entire study region. Figure 2.2 illustrates how a frame on a smaller-scale, areawide map readily indicates the area covered by a detail inset.

Larger-scale maps focused on a comparatively complex subregion might usefully include a *locator inset map*, as in Figure 2.3, to show at a much smaller scale the subregion's location within the overall study area. Although a separate, larger-scale map of Paris and vicinity most likely would not require a locator inset, readers with only a vague sense of the geography of France could benefit from a locator inset on a subregional map of Bordeaux. Unfamiliar or vaguely defined subregions or study areas almost always require a locator inset, unless the author uses a

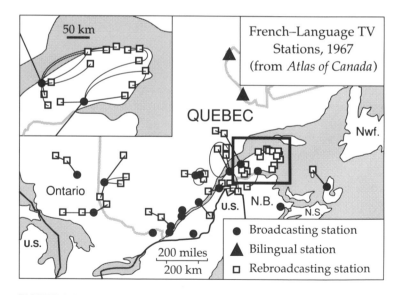

FIGURE 2.2. Frame on a smaller-scale, areawide map identifies the area covered by the detail inset map at the upper left.

separate, relatively detailed *locator map* to give the reader a geographic frame of reference. An anthropologist, archaeologist, or historian focusing an entire book or article on a comparatively small area might want to introduce the study with a separate locator map describing the study area's situation relative to familiar national or continental boundaries, important physical features, and important cities, kingdoms, culture areas, or trade routes.

Locator maps can be particularly useful if the map author decides to gain space by abandoning the traditional north-at-the-top orientation. This alternative can be advantageous for a study area with a pronounced northwest–southeast or northeast–southwest elongation, although an unfamiliar orientation might puzzle or mildly annoy a few readers. A north arrow can both signal and describe the map's atypical orientation, and a north-oriented locator inset showing the area covered by the larger-scale main map might make the unfamiliar orientation both understandable and acceptable.

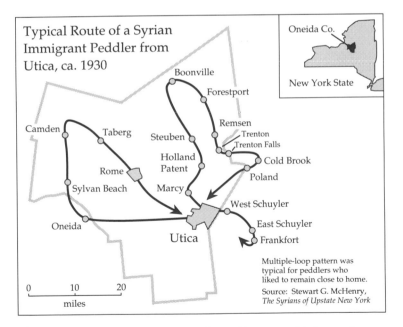

FIGURE 2.3. Locator inset map provides a wider geographic frame of reference for the area shown on the larger-scale main map.

Most readers will appreciate a *graphic scale,* which describes
one or more distances with a horizontal line and perhaps a few
tick-marks. After all, a half-inch line labeled "10 miles" is much
more readily related to both the world of the viewer and the symbols on the map than the corresponding approximate ratio scale
1:1,250,000. Graphic scales need not be highly detailed, unless
viewers are likely to be concerned about both relatively long and
relatively short distances. Having all graphic scales represent a
constant distance is an efficient signal to the viewer that scale
and degree of generalization vary. If the maps in a series differ
substantially in scale, comparisons of scale and relative size
might require graphic scales with overlapping distances. For example, scale bars on a nested series of maps of France might portray distances of 1/2 mile and 1 mile on a map of the Montmartre
district; 1 mile through 5 miles on a map of central Paris; 5, 10,
25, and 50 miles on a map of Paris and vicinity; and 50 and 100
miles on a page-size map of France. Because the scholar's audi-

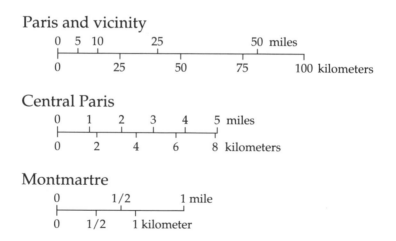

FIGURE 2.4. Overlapping scale bars for a nested set of four page-size maps
of Montmartre, central Paris, Paris and vicinity, and France as a whole include
English and metric units.

ence is often international, the scale bar might represent round-number distances in both kilometers and miles (or feet), perhaps with one type of unit above and the other below the scale bar, as in Figure 2.4.

Using a graphic scale can be the only safe strategy in case an editor, graphic designer, or printer decides to alter the size of a map, with or without the author's consent. I say "safe" because use of a ratio or fractional scale might lead to embarrassment or the added cost and delay of revising the illustration. For instance, reducing an 8-inch-wide 1:1,000,000-scale map to a width of 5 inches would invalidate a ratio or fractional scale printed on the map, because the scale of the reduced illustration would be 1:1,600,000, not 1:1,000,000. Yet a graphic scale labeled "20 km" still represents 20 km, because photographic or electronic reduction decreases its length proportionately, from 2 cm to 1.25 cm. As a general rule, use ratio or fractional scales with extreme caution when you cannot control or predict exactly the final publication size of the map.

Graphic scales are not always appropriate. Because they invite comparison with distances between map symbols, graphic scales should not be used on maps covering the whole world, an entire continent, or a large country, such as Canada or Russia. As the next section explains, the scale varies significantly on a map that must stretch and compress a substantial part of a three-dimensional, spherical surface to fit a two-dimensional sheet of paper. A map viewer who estimates the distance between Moscow and Oslo, for instance, using a graphic scale that is valid only along the equator, might overestimate this distance by a factor of two or more. It is important that a map author understands the basic principles of map projection. An ignorant map author might make embarrassingly inappropriate distance estimates or unwittingly introduce a wholly unsuitable distortion of routes, boundaries, and relative sizes.

GLOBAL PERSPECTIVE AND THE DISTORTION OF SIZE AND SHAPE

Among the flat map's many advantages is its complete, at-a-glance view of the world that a reader can scan rapidly, without reaching out to turn a bulky, expensive globe. Like most abstract representations, though, world maps trade precision for convenience. Because the geometric distortion required to project a curved surface onto a plane is much more severe and dramatic for small-scale world maps than for larger-scale maps of neighborhoods or even of large countries, the world map is a good place to begin to understand how map projections not only cause information to be lost, but also may inadvertently twist or contradict the map author's intended meaning.

Although the previous section examined scale as a uniform property of the map affecting selection of features and loss of detail, map scale actually is a function of both location and direction. That is, scale generally varies not only from place to place across the map but with direction as well. Thus, map scales around Helsinki will tend to differ from map scales around Rome, and the scale along a parallel running east–west through Helsinki might differ markedly from the scale along a meridian running north–south through the same point. Figure 2.5 illustrates this directional variation of scale with a world map based on the equirectangular projection, which stretches the north and south poles—mere points on the globe—into lines as long as the equator. In this example, north–south scale is constant across the map, but east–west scale equals north–south scale only at the equator. Compared to the equator, east–west scale is 1.34 times greater at Rome and 2.01 times greater at Helsinki. And at the poles, where a finite length represents a zero distance, east–west scale is infinitely large.

Quite obviously, the map in Figure 2.5 distorts several geographic properties, most notably the distances between points, the areas and shapes of continents and large countries, and the angles between meridians meeting at the poles. For example, the map exaggerates Greenland's relative area and east–west extent. In contrast, a globe would show that Greenland is less than an eighth the size of South America and has a pronounced north–south elongation. In general, map projections distort distances, angles, areas, directions, and the gross shapes of continents and large countries.

Some types of distortion can be controlled and even eliminated, whereas other types can only be minimized for selected parts of the map. For instance, all two-dimensional maps must distort noticeably the three-dimensional shapes of continental outlines and large countries, but a group of map projections called *conformal* can in theory preserve angles and the shapes of small circles and other local features.[1] For this reason, most large-scale maps, including most modern topographic maps, use projections that are conformal. After all, a city block that is a perfect square should look like a square on the map, and a traffic circle should look like a circle. Because large-scale maps generally cover relatively small areas in comparatively great detail, there is no perceptible distortion of distances, shapes, areas, or directions when the map projection is conformal. Although all these types of distortion are present, for the very small part of the Earth represented at a large scale on a desk-size sheet of paper, scale variation is a minor consideration compared to the map maker's ability to measure precise location on the Earth and to position symbols precisely on the map. Another, potentially more severe, source of distortion on a large-scale conformal map is shrinking and swelling of the paper caused by changes in humidity.

On a world map, in contrast, the map maker will want to preserve the true relative areas of continents, regions, countries, and provinces. Map projections that preserve area relationships are called *equivalent*, or simply *equal-area*. Conservation of area is particularly important for maps viewed by school children, who might otherwise develop a distorted notion of the relative sizes of continents. When my daughter was in seventh grade, I visited her school on the annual Parents' Night. I was at first amused and then shocked to find that the only wall map of the world in her

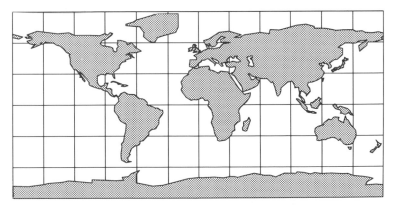

FIGURE 2.5. Equirectangular projection of the world illustrates an extreme directional variation of scale. (This and other examples of map projections used in the chapter were created with WORLD, a map projection software package developed at the University of Minnesota.)

social studies classroom had a Mercator projection, which among other distortions enlarges Greenland to the size of South America. An excellent tool for navigators and those studying navigation, a Mercator chart has no place as a general reference map.

Preservation of area relationships is also important for dot-distribution maps, on which individual dots represent a stated number of, for example, people, sheep, automobiles, bushels of wheat, or houses of worship. In a region of intensive sheep production, for instance, the dots will be relatively close, perhaps even beginning to coalesce, whereas in a region where raising sheep is less intensive, the dots will be more widely separated. Because the purpose of a dot-distribution map is to show geographic variations in density, the map will be correct only if dots are plotted on an equal-area base map. Figure 2.6, a schematic illustration of the effects of area distortion on an agricultural economist's map of sheep production, shows three small, square-shaped study regions in southern Arabia, southern Argentina, and central India. On the globe all three regions have the same size, but the plots in Argentina and India each have

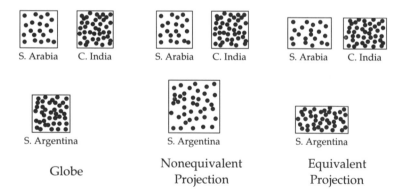

FIGURE 2.6. Hypothetical example of how a nonequivalent projection (center) might distort dot density, whereas both a globe (left) and an equivalent projection (right) would preserve density relationships by portraying relative area correctly. The equivalent projection (right) significantly distorts shape in order not to distort area. Hypothetical square areas in southern Arabia and central India, at similar latitudes, are distorted less than an equal-size area in southern Argentina, which is further from the projection's line of contact at the equator with a globe of the same scale.

eight million sheep, whereas the plot in southern Arabia has only four million sheep. If the economist represents 200,000 head of sheep with a dot, the equally dense regions in Argentina and India will each have forty dots and the less dense region in Arabia will have only twenty dots. The left side of Figure 2.6 shows how a globe would portray correct relative densities and shapes, the center demonstrates the effect of a map projection that progressively exaggerates area with increased distance from the equator, and the right side shows the correct relative densities on an equal-area map. Because southern Arabia and central India are at roughly the same latitude, the center map does not noticeably distort size and density. But because southern Argentina is much farther from the equator, the map doubles its projected area and markedly reduces its density. By distributing forty dots over an area twice as large, the map suggests incorrectly that southern Argentina raises sheep no more intensively than southern Arabia.

Designing or selecting a map projection involves many tradeoffs. Particularly prominent is the penalty incurred in preserving relative area at the expense of markedly greater distortions of shape. Look, for instance, at the equirectangular world map in Figure 2.5, and consider how a cartographer might make this projection equivalent. One very simple way of preserving relative area is to adjust the spacing of neighboring parallels to reflect relative areas on the globe. For instance, because the zone between 60° N and 90° N contains roughly 1/15 of the world's area, not the exaggerated 1/6 that results from a constant north–south scale, these two parallels should be closer. In the eighteenth century, the mathematician Johann Heinrich Lambert (1728–77) discovered how to do this by making north–south scale a trigonometric function of latitude. But as Lambert's cylindrical equal-area projection in Figure 2.7 illustrates, preserving area at the expense of conserving shape yields a severe north–south compression of poleward features.

Equivalence is not the only visual asset of Lambert's projection. Its *cylindrical* grid of perpendicularly intersecting sets of straight-line parallels and straight-line meridians resembles the rectangular grid of most large-scale maps and makes it easy to estimate latitude and longitude or portray time zones. The label "cylindrical" reflects the conceptual mathematical development of rectangular projections as a two-step process, first to shrink the Earth to a globe, and then to project onto a surrounding cylinder in contact with the globe at the equator. The family of cylindrical projections includes the equirectangular projection in Figure 2.5, which also has the characteristic straight-line grid.

Pseudocylindrical projections carry this two-stage process a step further by bending the meridians inward toward the poles. The result is an additional tradeoff between an easy-to-plot grid and a better representation of shape for selected continents or countries. The sinusoidal projection in Figure 2.8 is particularly useful for conserving the shapes of tropical countries and areas near its *central meridian*, the straight line connecting the poles. When centered conventionally at the Greenwich meridian, as in Figure 2.8, the sinusoidal projection provides a low-distortion

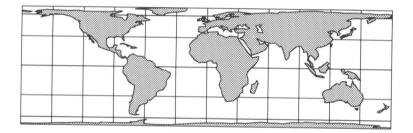

FIGURE 2.7. Lambert's cylindrical equal-area projection.

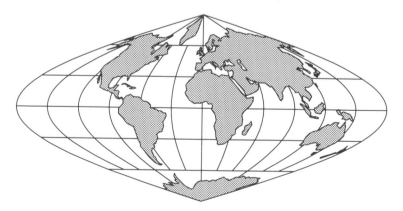

FIGURE 2.8. The sinusoidal projection.

view of Europe and Africa—but only at the cost of severe distortion in northeast Asia, Alaska, Antarctica, and other peripheral areas. Because the central meridian can be moved, where the map author "centers" this projection becomes an important design decision; a good representation of some regions of the world is traded for gross distortions of shape elsewhere.

In addition to equivalence and converging meridians, the sinusoidal projection has other noteworthy attributes. Its name reflects a trigonometric foundation: its meridians are sine curves, evenly spaced along the parallels, which remain straight and parallel. Evenly spaced parallels provide an accurate representation of latitudinal relationships, particularly important for geographic distributions influenced by climate. The rapid, somewhat unrealistic convergence of the meridians toward the poles reflects the preservation of east–west scale along the parallels.

Severe angular distortion in areas far from both the central meridian and the equator, and its overall Christmas-ornament shape, make the sinusoidal projection less suitable for a world map than the more oval Mollweide projection in Figure 2.9. Using more gently rounded ellipses instead of sine curves as meridians, the Mollweide has a pleasing oval shape more suitable for a global map. These elliptical meridians require decreasing the north–south scale near the poles and increasing it near the equator, so that the lowest distortion occurs in the middle latitudes. In Figure 2.9, North Africa, the Mediterranean area, and the southern tip of Africa have the least distortion, while Alaska, north-

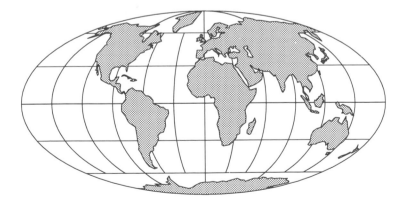

FIGURE 2.9. The Mollweide projection.

east Asia, and Antarctica still reflect the maximal distortion of upper-latitude peripheral areas. But by sacrificing evenly spaced parallels, the Mollweide projection preserves area relationships without distorting these fringe areas as greatly as the sinusoidal projection.

According to the theory of map projections, equivalence and conformality are mutually exclusive properties: no map can preserve both area and angles. But there can be acceptable compromises between conserving area and conserving angles, as geographer Arthur Robinson recognized in a world map projection that the National Geographic Society adopted in 1988 for its general-purpose whole-world reference map.[2] Figure 2.10 illustrates some of these compromises. Evenly spaced meridians that curve without converging to a point don't distort peripheral shapes as greatly as the sinusoidal and Mollweide projections do. Straight-line parallels that are evenly spaced only between 38° N and 38° S still distort area near poles, but not as severely as the equirectangular projection. Greenland and Antarctica are bigger than they should be, but area distortion is minimal for regions with over 99 percent of the world's population. The Robinson projection's other name, *orthophanic*, which means "right appearing," reflects its generally realistic portrayal of familiar shapes.

Figure 2.11 uses an ellipse-shaped point symbol called *Tissot's indicatrix* to show the spatial pattern of distortion on the Robinson projection.[3] Each symbol represents the projection's distortion of a very small circle at a particular location on the map. The elongation or eccentricity of the symbol represents the degree of angular distortion, and its relative size reflects areal distortion. On a conformal projection the indicatrix would be circular everywhere, yet vary in size to reflect differences in scale from point to point. In contrast, for an equal-area projection, elliptical-shaped indicatrixes would vary widely in elongation but not in area. As the absence of any perfect circles in Figure 2.11 suggests, no point on the Robinson projection is completely free of distortion. But the projection's comparatively low distortion for the middle latitudes and the tropics outweighs its comparatively high distortion near the poles. This example demonstrates that the indicatrix plot can be a useful design tool for choosing the projection for a world or regional map.

Map authors who require an equal-area world map projection might consider interrupting the world map over the oceans in order to portray the continents with greater accuracy. By pointing out the geographic tradeoffs between low distortion near the central meridian and high distortion in peripheral areas, the sinusoidal and Mollweide projections suggest the possibility of a composite projection, developed by dividing the world into lobes interrupted over water and assigning each lobe its own locally centered pseudocylindrical projection. University of Chica-

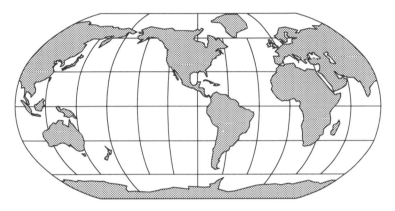

FIGURE 2.10. The Robinson projection, centered on 90°W to favor North America.

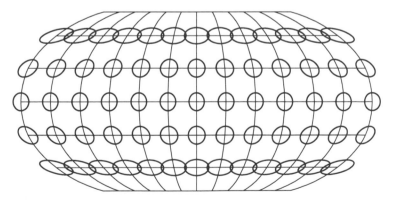

FIGURE 2.11. Tissot's indicatrix, plotted at grid intersections, shows the Robinson projection's pattern of angular and areal distortion.

go professor J. Paul Goode (1862–1932), who developed the widely used equal-area projection that bears his name, went one step further. Goode not only divided the world into the six lobes shown in Figure 2.12, but divided each lobe into two sections.[4] Between the equator and 40°44' the projection is sinusoidal, with evenly spaced parallels. Beyond 40°44' Goode used the Mollweide projection, which provides a truer representation of shape in the middle and upper latitudes by progressively reducing the spacing of parallels toward the poles. The two projections for each lobe share a common central meridian, selected to minimize the distortion of angles and shape. Called "Goode's homolosine projection" because it merges the homolographic, or Mollweide, projection with the sinusoidal projection, this *interrupted* projection commonly repeats parts of Siberia and Greenland so that these areas are intact on at least one lobe.

The pronounced east–west elongation of both Goode's homolosine projection and the rectangular equal-area projection in Figure 2.7 can severely limit the scale and detail of a horizontally oriented world map in a book or journal. Even a full-page, sideways, vertical layout might not be adequate. Rather than extend the map across two pages, with a break for the gutter, the map author concerned exclusively with land features can condense the projection by removing most of the Atlantic Ocean and then joining the remaining pieces together, in order to fit the map onto a stingy type page. Potentially useful for both interrupted and noninterrupted projections, condensing is a strategy that sacrifices completeness and familiarity to achieve clarity.

As the examples examined thus far demonstrate, whole-world maps always distort some areas more than others. Given the impossibility of a geographically egalitarian flat map, the map author should consider carefully the value of "recentering" the projection to favor a particular part of the world more relevant to the audience or to the study than Europe or Africa. As Figure 2.10 demonstrates, a projection centered on a North American meridian such as 90° W, which lies between Chicago and Minneapolis, is generally more suitable than a projection centered on Greenwich for a world map addressing a largely American audience. If the theme of the map is American foreign policy, for instance, this type of recentering might promote a better balance of cartographic symbols and support a narrative using the conventional hemispherical metaphors "east" and "west." In contrast, a political scientist concerned with Japan's foreign policy might prefer a projection centered on 140° E, a meridian passing through Tokyo's eastern suburbs. Figure 2.13's Japan-centered world interrupted at 40° W presents a far more appropriate cartographic stage than, say, a Greenwich-centered projection, which would hide Japan in the northeast corner and grossly distort its shape as well. By simplifying the tedious calculations and plotting required, computer graphics and map projection software allow the map author to explore the merits of various views and to choose a projection tailored to the topic. (As discussed in

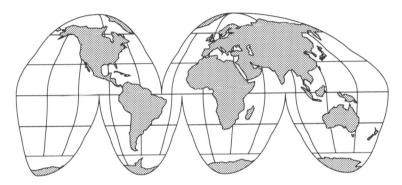

FIGURE 2.12. Goode's homolosine projection.

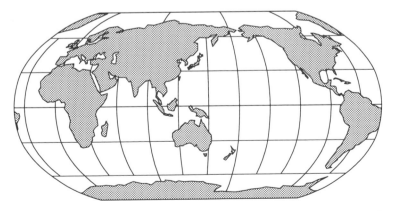

FIGURE 2.13. The Robinson projection centered on 140° E and interrupted symmetrically at 40° W to favor Japan.

chapter 5 and appendix B, map librarians and experienced cartographic illustrators can be helpful in finding a base map with an
appropriate projection.)

REGIONAL PERSPECTIVES AND THE CONSERVATION OF DISTANCE

Equirectangular, sinusoidal, Mollweide, and Goode's homolosine projections, and the strategies of condensing and recentering, by no means exhaust the map author's choices. This section
extends the examination of whole-world maps to continental,
national, and provincial maps. At the somewhat more detailed
scales possible with smaller portions of the earth, the tradeoff
between preserving area and conserving angles and small shapes
is no longer paramount. After all, as the map's geographic scope
decreases, a well-chosen projection can more easily accommodate both shape and area. Moreover, because the scholar examining a nation or region often must compare distances, minimizing the distortion of distance becomes an important goal.

   Continental, national, and regional map projections often re-

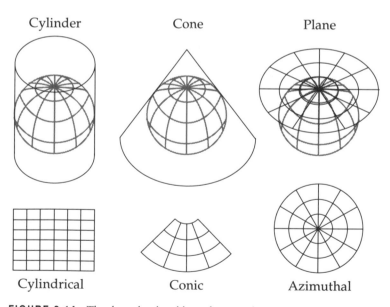

FIGURE 2.14. The three developable surfaces used in map projection
(above) yield distinctly different graticules (below).

quire one of the two other flattenable surfaces available to the
map author, the cone and the plane. As Figure 2.14 demonstrates, each of the three *developable surfaces* has a distinctive
grid, or *graticule*, in its normal, pole-at-the-top orientation: the
classic cylindrical projection has straight-line meridians and parallels; a conic projection has converging straight-line meridians
and concentric, partial-circle parallels; and an azimuthal projection formed by projecting to a plane touching the globe at either
pole has straight-line meridians converging to the common center of its full-circle parallels. In these most straightforward cases,
the developable surface just touches, or is *tangent* to, the globe at
a single circle or point. Cylindrical projections typically are tangent at the equator, conic projections at a middle-latitude parallel, and azimuthal projections at one of the poles.

Distortion, which is lowest where the developable surface is in
contact with the globe, grows with increased distance from the
tangent point or circle. Consequently, cartographers recommend
cylindrical projections for continents such as Africa or South
America, which straddle the equator; conic projections for midlatitude continents, such as Asia, Australia, Europe, and North
America, which can straddle a carefully specified *standard parallel*; and azimuthal projections for Antarctica and the northern
polar region. This rule also helps the map author select a low-
distortion projection for smaller areas. Cylindrical projections
provide good maps of Indonesia and other tropical countries, for
instance, whereas conic projections favor more poleward nations such as China and the United States. The latitude principle
for selecting a developable surface also yields maps that look
right in the sense of resembling the globe's portrayal of the region
in question.

The map author seriously concerned with reducing distortion
has several other available strategies. The most obvious option is
to add a second line of contact, by making the developable surface pierce the globe rather than merely touch it. The resulting
secant projection has two standard parallels if it is cylindrical or
conic, or a standard parallel instead of a tangent point if it is azimuthal. As Figure 2.15 illustrates, features on a secant projection
are closer on the average to any standard line—and thus less distorted—than features on a tangent projection. Because carefully
chosen standard parallels can ensure comparatively low distortion throughout the study region, the map author might want to

experiment by plotting Tissot's indicatrix, as in Figure 2.11, or to follow the practices of government mapping agencies. Standard parallels of 29°30' and 45°30' N for the Albers equal-area conic projection and of 33° and 45° N for the Lambert conformal conic projection provide a nearly optimal portrayal of the forty-eight conterminous United States (Figure 2.16), if the map author wants to preserve either area or small shapes. Secant conic projections with two standard parallels are especially useful for low-distortion representations of regions, countries, states, or provinces with a marked east–west elongation, such as Switzerland, the United States, and Tennessee.

For regions with a north–south prolongation, such as Chile, Norway, and Mississippi, a low-distortion map usually requires a *transverse cylindrical* projection. Orienting the cylinder so that

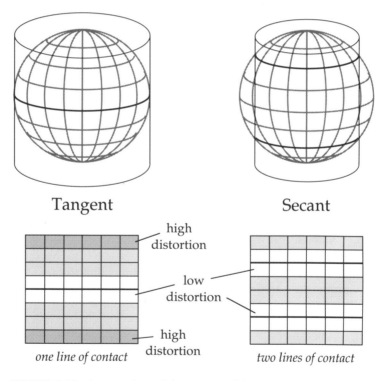

FIGURE 2.15. A comparison of the patterns of distortion for tangent (left) and secant (right) cylindrical projections.

Albers equal-area conic projection

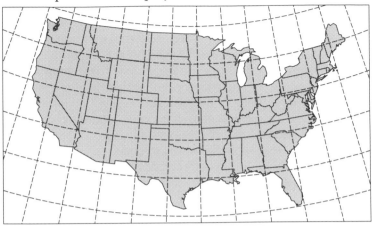

Lambert conformal conic projection

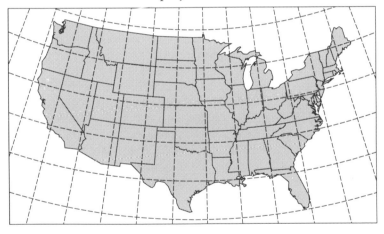

FIGURE 2.16. Two common conic projections of the 48 conterminous United States. That these representations seem virtually identical attests to the low distortion of angles within the area shown on the Albers equal-area conic projection (top), which does not distort area at all, and the low distortion of area on the Lambert conformal conic projection (bottom), which does not distort angles. Extension of the graticule beyond the 48 states would reveal notable differences between these two projections.

its axis pierces the globe at diametrically opposite points on the equator yields a low-distortion zone that runs north–south.[5]

Azimuthal projections are common in a variety of aspects: polar, equatorial, and oblique. Projection planes centered over either pole offer a correct-looking view of the Antarctic, as well as a realistic portrayal of the efficiency of near-polar airline routes and an often-forgotten yet geopolitically significant view of the proximity of North America and Russia.[6] By focusing on the Americas, say, a projection plane centered at a point along the equator affords another geopolitically significant view, useful perhaps for what it omits as well as what it shows. Equatorially centered hemispheric projections can be useful for map authors addressing U.S. trade with Latin America, European exploration and exploitation of Africa, and a variety of anthropological, foreign-trade, geopolitical, and historical questions concerned with the enormous Pacific Basin. By focusing the reader's attention on a particular city or country (such as Seoul, Korea, in Figure 2.17), an oblique azimuthal projection gives the place at its center the lead role on a hemispheric or even global cartographic stage. As news maps sometimes demonstrate, the oblique azimuthal projection also provides an intriguing global perspective for locator insets on maps of foreign areas.

An azimuthal projection can favor its center point in several ways. All straight lines through the center represent shortest-distance routes along great circles, and the angles between these lines are correct as well. Indeed, the name "azimuthal" reflects the center's true portrayal of *azimuths*, that is, the relative directions of these straight-line great circles measured in degrees clockwise from north. At the expense of enormous distortions of area and shape near its periphery, an *azimuthal equidistant* projection also portrays true relative distance along these converging great-circle routes. Uses of the azimuthal equidistant projection include maps showing the spread of an idea, a religion, a language, a disease, or any other widespread phenomenon that originated in a small part of the world. The azimuthal equidistant projection in Figure 2.17 could be particularly useful for examining, for instance, the relationship between various nations' per capita purchases of Korean-made consumer products and those nations' distance from Seoul.

Another azimuthal projection might be useful for maps showing direct long-distance air routes. The *gnomonic* projection has the unique property that a straight line between any pair of points on the map is a great-circle route. On a gnomonic projection, for instance, an Alaskan stopover on a flight from Washington to Beijing would appear natural; in contrast, it is represented as a dog-leg detour on cylindrical projections. Neither equivalent nor conformal, the gnomonic projection favors its center principally with relatively less outrageous distortions of area, distance, angles, and gross shape. But as Figure 2.18 illustrates, distortion increases so rapidly with increasing distance from the center that on a gnomonic projection centered at one of the poles the equator would lie at infinity.

In navigation, the gnomonic projection is sometimes used with the Mercator chart, which I am largely ignoring here because humanists and social scientists are seldom concerned with the mechanics of navigation.[7] A marvelous invention for sailors, because it shows lines of constant direction, called *rhumb lines*, as straight lines, the equatorially centered Mercator projection is

FIGURE 2.17.   An azimuthal equidistant projection centered on Seoul, Korea.

a cylindrical projection that, as Figure 2.19 demonstrates, so grossly distorts area and distances that the poles lie off the map at infinity. For centuries, publishers and social studies teachers, who should know better, have used it as a standard base map of the world for atlases and wall maps. Historians portraying the routes of fifteenth- and sixteenth-century explorers should already be well aware of the Mercator chart and its correct use and should require no further introduction. And the rest of us need only to avoid it like the cartographic equivalent of rabies.

For scholars concerned with neither navigation nor the directness of long-distance routes, the gnomonic projection might be noteworthy principally as a special case of the general vertical-perspective projection. The point of perspective can be above the earth, on the opposite side, or in the case of the gnomonic projection, at the center. But if the vantage point is above the earth, as in Figure 2.20, the perspective azimuthal projection offers a dramatic Earth-from-space view similar in geometry to a snapshot of the globe. Including meridians and parallels, at least over wa-

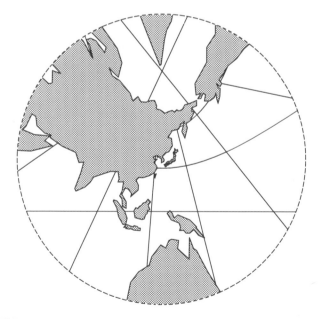

FIGURE 2.18.   A gnomonic projection centered on Seoul.

ter, not only enhances the pictorial effect but also warns the viewer of severe distortion near the periphery. If distance relationships are of minor importance, an oblique-perspective projection can yield an interesting locator map or an attractive cover illustration. Modern map projection software allows the map author to experiment with both the projection's center and the height of its point of perspective.

Another modified azimuthal projection, the tilted-perspective projection, provides even more dramatic, unevenly distorted views of the earth, as Figure 2.21 demonstrates. Illustrators and art directors appreciate not only the shock value of its unfamiliar orientation but also its ability to cover a wide area while focusing on a detailed foreground. The widely imitated cartographic car-

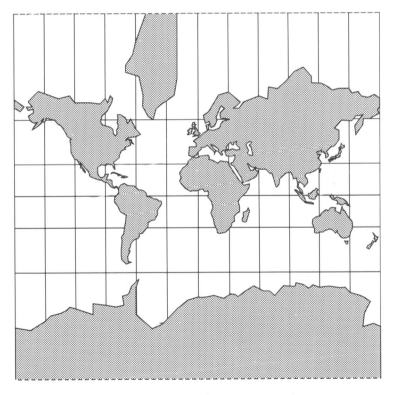

FIGURE 2.19. An equatorially centered Mercator projection.

toon showing a highly provincial "New Yorker's View of the United States" conveys the idea of the tilted-perspective projection. Despite its association with geographic levity, the tilted-perspective projection warrants the attention of the serious scholar who wants to focus the viewer's attention on a directional trend, such as the concentration of French-speaking households near Maine's border with Quebec or the generally westward movement of the settlement frontier through the Middle Atlantic states during the sixteenth and seventeenth centuries. Map projection software allows the map author to manipulate a hypothetical camera by specifying the latitude and longitude of a location, the height of a hypothetical camera at that location, the photographer's viewing rotation in degrees clockwise from north, and the tilt of the camera's axis upward from a perfectly vertical, downward view.[8] The higher the point of perspective above the surface, the larger the area covered on the map. So-

FIGURE 2.20. An oblique vertical-perspective projection with its center at Paris, 48°51' N, 2°02' E, and its point of perspective 5,000 miles (8,065 km) above the Earth's surface. The jagged coastlines of generalized polygons representing land masses are more apparent at this larger scale than in previous examples in this chapter.

phisticated computer graphics systems promote the camera metaphor further by adding a zoom lens, three-dimensional terrain, and colors representing land cover.

College students often find map projections more perplexing and frustrating than any other topic in an introductory cartography course. Especially if the treatment is at all mathematical, map projections can be difficult, to be sure. Before computer graphics, laboratory assignments could be computationally tedious as well. Some instructors cover projections early, to clear the way for more interesting topics, whereas others move it toward the end, perhaps to satisfy a vague pedagogic notion that course content should become progressively more difficult, not less so. These essays on expository cartography clearly warrant the former approach, because projections are basic to understanding and using maps. In steering clear of the underlying mathematics and the rich variety and fascinating history of map projections, I have focused on the types of distortion, the types of developable surface, and the simple rule that distortion increases with distance from a line or point of contact or from a central

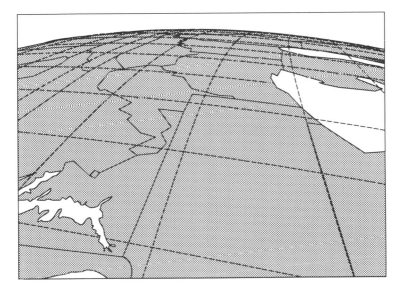

FIGURE 2.21. A tilted-perspective projection centered above New York City, but looking westward and downward toward southwestern Pennsylvania.

meridian. And I have tried to point out why the map author must not select a projection solely because it is readily available, it looks interesting, the software manual seems to favor it, or other scholars have used it.

SOME GENERAL RULES

Before moving on to map symbols and visual variables, I offer five basic rules to guide the inexperienced map author hesitant about selecting a map projection.

First, determine whether the map must conserve a general geometric property such as angles, area, or distance. Large-scale maps, which should distort neither angles nor the shapes of small objects, must have a conformal projection. Small-scale dot-distribution maps, particularly if they cover the world, a continent, or a large country, should have an equivalent projection, as should maps on which the viewer might be required to compare areas. If the viewer must compare distances to or from a designated focal point, the projection should be equidistant.

Second, if a map-projection software package is not available, apply the latitude rule to select a developable surface. For polar areas, choose an azimuthal projection. For midlatitude regions, choose a conic projection. For tropical regions, choose a cylindrical projection. And for a world map, choose a low-distortion projection, such as the Robinson projection, if both water and land are important and if some distortion of area can be traded for a balanced representation of area and shape. If the map can focus on land and the oceans are unimportant, Goode's homolosine equal-area projection usually is a safe choice.

Third, if a computer with projection software and an appropriate cartographic database is available, consider the benefits of recentering the projection to minimize distortion in the more important parts of the map. Recentering might make compilation easier, if the map projection can be tailored to match a standardized projection common to the map author's principal cartographic sources. And recentering can be essential if available base maps interrupt the globe awkwardly. For instance, a projection centered on Greenwich and interrupted in the Pacific would be a poor choice for a world map concerned with the foreign affairs of Asian nations. Recentering and reorienting the projection might also focus attention on a dominant trend or distance from an influential center. Ready recentering also allows the map author to ignore the latitude rule or to address situations in which the geographic scope of the map is not easily pigeonholed.

Fourth, consider other relevant factors or special requirements, such as the shape of the region in question or the need for a large map key or block of explanatory text. If climate or another comparison based on latitude is important, a projection with evenly spaced, straight-line parallels is warranted. If differences in time of day are important, straight-line meridians are appropriate. If the map is highly detailed and an efficient use of available space is essential, unfamiliar projections or orientations might be helpful. For instance, an oblique-perspective projection that places an area with markedly more detail in the foreground can be useful if all viewers are likely to know the region and if distance relationships and shapes are much less important than relative proximity. Oblique views can be suitable for maps of neighborhoods, cities, states, and provinces, as well as maps of countries and continents. Because scale varies widely from place to place and with direction, the map author must not include a scale bar.

Fifth, and above all, remember that flat maps can significantly distort large portions of Earth. As a general rule, omit bar scales on all maps spanning a thousand miles or more. Avoid such a demonstrably bad choice as a Mercator chart for a world map not related to navigation. When transferring features from one projection to another, note carefully the types of distortion and their geographic patterns. If a single map projection cannot satisfy the diverse needs of the data or the concepts discussed, consider two views, either juxtaposed for ready comparison or separated for proximity to the associated text. Like clear sentences and well-structured paragraphs, effective maps require material tailored to a single, well-defined topic or idea.

Visual Variables
and Cartographic
Symbols

AS A SCALED-DOWN REPRESENTATION OF REALITY, the flattened, two-dimensional map projection is a stage for text and graphic symbols describing boundaries, places, and other selected features. Text is necessary in a map's title, key, and source notes. When positioned appropriately, text can identify simple point symbols as specific cities and designate some lines as highways and others as railways. Yet text is often of limited help to the viewer who needs to differentiate rapidly among various classes of features on a map crowded with symbols and labels. Reading individual labels takes time, after all, and redundant words or abbreviations consume space otherwise useful for showing additional features. A superior strategy is to code the points, lines, or bounded areas representing the position and extent of features to describe as well their general character or significance.

Graphic coding might also differentiate among planes or layers of information. For instance, if a reference map's line symbols incorporate carefully developed and consistently applied graphic codes, the viewer can not only distinguish readily between railways and freeways but also efficiently identify the structure of a particular network. On a map designed to communicate a geographic pattern or relationship, carefully orchestrated symbols can focus the viewer's attention on one or two more important, foreground planes of information, while making available several background planes of information to provide a geographic frame of reference.

Specific types of cartographic symbols have a functional association with specific types of data. French cartographer-semiologist Jacques Bertin used this functional relationship to develop a theory of cartographic communication based on eight "visual variables."[1] Understanding Bertin's visual variables and their functions can help the map author select an effective scheme from the wide and often confusing array of available symbols, whereas violating these functional associations encourages confusion and misinterpretation. After describing the visual variables and their appropriate roles in cartographic communication, this chapter examines several basic cartographic functions or objectives, such as portrayal of a route or of a location and its neighbors, each of which calls for a focused application of one or two visual variables. These basic objectives collectively encompass many of the simpler maps used to illustrate scholarly writing in the humanities and social sciences.

The Visual Variables of Jacques Bertin

Bertin recognized two kinds of visual variables: retinal variables and locational variables. His two locational variables are the familiar horizontal and vertical coordinate axes of the scatterplot and other two-dimensional data graphs; for maps based on a rectangular projection these axes correspond to longitude and latitude. Bertin's interests were broader than cartography, and his schema address diagrams and network charts, as well as maps of various types. Cartographic treatments of Bertin's visual variables commonly ignore the locational variables, which seem more a part of map projections than a topic in symbolization. Yet, as I observe later, particularly in chapter 8, the map author often needs to supplement maps with statistical charts; that is, some geographic relationships are most thoroughly understood when portrayed in both the "geographic space" of the map and the "attribute space" of the data graph.

By separating his six retinal variables from the two locational variables, Bertin drew a clear distinction between the spatial relationships among the graphic symbols on a map and the perceptual properties of the symbols themselves. The six arms of Figure 3.1 represent six unique ways of differentiating graphic marks, and thus six different ways of using graphic symbols to represent differences among features or places. In Bertin's conceptualization, the locational variables first fix a graphic mark at a point in the plane, and the retinal variables then "elevate" that mark above the graphic plane by presenting the eye with a pattern of light different from that of the background or nonprinted paper.

The names of the retinal variables require little explanation. *Shape*, *size*, and *orientation* are fundamentally self-explanatory. *Texture* refers to the size or spacing of smaller, repeating elements of a symbol, varying from fine to coarse. *Hue* is that aspect of color associated with wavelength in the comparatively tiny portion of the electromagnetic spectrum visible to the human eye. Because this book has no color plates, Figure 3.1 identifies by name a sampling of five readily distinguished hues. *Value* is the term most likely to confuse readers not familiar with printing technology and color theory. When artwork and text are printed using only black ink, "value" refers to a symbol's graytone value on a scale running from black to white through various intermediate shades of gray. Because "value" refers indirectly to ink density or the relative area covered by fine-grained dots or lines, a set

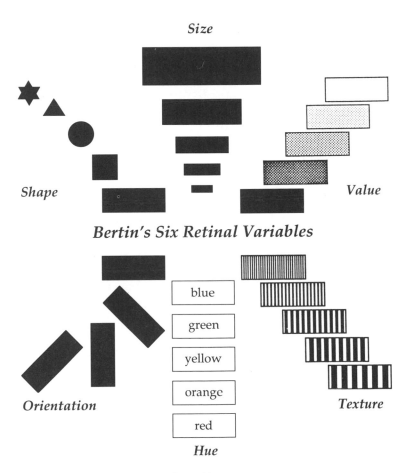

FIGURE 3.1. Bertin's six retinal variables.

of symbols printed only in red ink could vary on a value scale running from solid red to white through various intermediate shades of light red and pink.

Figure 3.1 reveals a functional specialization among the retinal variables. Size, value, and texture yield graphic marks that the eye readily organizes as a quantitative dimension, with lower numbers or ranks at one end and higher numbers or ranks at the other. These quantitative retinal variables can be efficient on maps showing an ordering of places or features ranked by differences in amount or intensity. In contrast, shape and hue provide a qualitative differentiation among features varying in type or kind. Because each shape and hue looks different, symbols differing in shape or hue are ideal for portraying categorical distinctions among geographic features. And varying the orientation of the graphic mark is the most direct and logical way of showing the orientation of the feature itself. Orientation is an important visual variable for symbols that represent features with an identifiable direction, such as lines portraying roads or rivers or arrows showing winds or ocean currents.

The quantitative retinal variables have an important secondary specialization: size is ideal for portraying amount or magnitude, whereas value is more suited to portraying relative amount or intensity. Thus, a map showing the respective populations of various cities would employ symbols varying in size, whereas a county-unit map showing the percentage of each county's population residing in an urban area would use symbols varying in value. Two common metaphors describe the logic underlying these symbol-data associations. The link between symbol size and amount reflects a notion of "the bigger the greater, and the smaller the less." In contrast, the relationship between value and percentage reflects a sense of "the darker the more intense, and the lighter the less intense."[2] Intensity measurements for places or features represented logically by variations in value include means, medians, densities, ratios, rates of change, proportions, and percentages.[3] Texture, the third quantitative retinal variable, is useful largely as a low-resolution substitute for value and thus is more readily associated with intensity measurements than with amounts or counts.

Discussion of the retinal variables and their use in map design must consider the dimensionality of graphic symbols, of which there are three types: point symbols, line symbols, and area symbols.[4] Point symbols are positioned at points, which are geometric objects with locations but no dimensions. In contrast, line symbols represent one dimension, length, and area symbols represent two dimensions. Because the utility of retinal variables differs markedly with symbol dimensionality, this distinction provides a convenient framework for examining some fundamental decisions confronting the map author.

Cartographic features, like graphic symbols, can conveniently be categorized dimensionally: as point features, line features, area features, or surface features. A feature's category might depend upon map scale and the degree of generalization. Cities, for example, tend to be area features on large-scale maps but point features on small-scale maps. Although symbols and geographic features often share the same dimensionality, the map author can assign a symbol to a feature with a different number of dimensions. Thus, on many large-scale maps boundary lines, rather than area symbols, commonly differentiate among counties, cities, and minor civil divisions. And a map author who wants to show variation among counties in the number of library books circulated might position at each county's center a point symbol scaled according to size.

Point Symbols

Point symbols not only mark specific point locations, but often also describe one or more attributes of these locations. Figure 3.2 shows the principal retinal variables used for cartographic point symbols, which differ most commonly in size or shape. Symbols varying in size usually are uniform in shape, unless the map author needs to portray on one map the magnitude variations of two or more types of places or features. Symbols varying in shape can be either geometric or pictorial. Simple pictorial icons that represent features recognizably and unambiguously can spare the map viewer tedious references to the map key, especially on large maps crowded with point symbols representing a variety of feature categories. Orientation is the primary retinal variable for arrows representing direction of movement and is an important secondary variable for tiny icons representing military units, ships, or barriers. Color can be a useful redundant or secondary stimulus for differentiating among opposing military units represented by geometric and pictorial point symbols. But hue and graytone value tend not to function effectively as the primary retinal variable for dots and other tiny point symbols with surface areas too small to readily differentiate a symbol from its neighbors. Shape works better than hue, and size works better than value. As Figure 3.2 illustrates, simple point symbols such as dots can also vary in numerousness or countability, which is akin to a size variation that can be subdivided and spread around to represent density or areal extent.

Quantitative point symbols can be scaled to vary in length, area, or volume. As Figure 3.3 illustrates, linear scaling generates symbols that grow comparatively rapidly in height. In contrast, with area scaling a circle or square representing a value of 4 is only twice the height of a symbol representing a value of 1. And with volume scaling a sphere or cube representing 4 is not

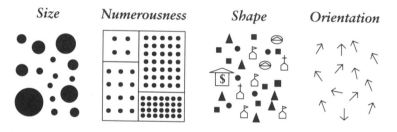

FIGURE 3.2. Common retinal variables for cartographic point symbols.

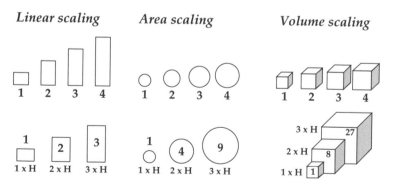

FIGURE 3.3. Comparison of point symbols scaled in one, two, and three dimensions. Top row illustrates progressively slower growth in height of symbols scaled by area and volume. Bottom row shows relative amounts represented by symbols one, two, and three units tall.

quite 1.6 times as tall as a symbol representing 1. The bottom row of Figure 3.3 demonstrates that symbols varying in height from 1 to 3 units can cover values ranging from 1 to 9 for area scaling and from 1 to 27 for volume scaling. Because area and volume scaling provide greater visual discrimination among smaller values, map authors may find these approaches attractive for mapping phenomena clustered toward the lower end of the range. Without training, though, the eye-brain system tends to underestimate the values of larger symbols, because a symbol's relative size is estimated as a crude mental average of its area and height.[5]

Graduated point symbols provide three kinds of information. The map viewer can (1) estimate values represented by particular symbols, (2) compare two symbols by estimating their relative size as a ratio or fraction, or (3) assess broad regional patterns portrayed collectively by all point symbols on the map. *Anchor stimuli*, representing a few specific labeled values in the map key, relate all the graduated point symbols to their corresponding data values. The representativeness of these anchor values affects the precision of viewer estimates of individual point symbols.[6] But representative anchor stimuli support only the first of these three information retrieval tasks, because the map viewer comparing two places or searching for regional trends has little need to consult the key.

In choosing a scaling method, the map author must consider how the range and dispersion of data values might affect the viewer's ability to judge differences in symbol size. A wide range with a few very large data values tends to reduce the precision with which viewers can recognize differences among smaller symbols. When the distribution is highly skewed by a few large values, or *outliers*, volume scaling can be useful for making differences more noticeable at the lower end of the scale. For many studies, after all, the difference between a town with 10,000 residents and a city with a population of 50,000 might well be as important as the difference between a major city with one million inhabitants and a metropolis with a population of five million. Area and volume scaling are less useful when data values are spread uniformly across a small range.

Cartographic symbols should serve the viewer's information requirements. If a quantitative map is intended to distinguish among only three or four categories of point phenomena, the map author might compress the continuous range of data values into a limited number of meaningful categories, with each higher category represented by a noticeably larger point symbol and its anchor stimulus associated with a range of data values. The resulting *range-graded point symbols* sacrifice a slightly more precise view of the raw data for a more reliable presentation of what the author deems significant. If pair comparisons are important, the map author might want either to avoid two- and three-dimensional scaling or to supplement the map with bar graphs comparing important magnitudes discussed in the text. If the viewer's comprehension of the overall geographic pattern of the distribution is more important than pair comparisons, area or volume scaling might be useful when small data values predominate.

Because map viewers can more reliably decode simple binary symbols than graduated point symbols representing a continuous range of values, two-category point symbols can be small and numerous. The left side of Figure 3.4 presents several examples of how a point symbol's interior can carry a straightforward two- or three-category code. Outlined point symbols with variable solid or vacant interiors promote ready comprehension through a straightforward presence-absence metaphor. At the simplest level, a small circle shows the location of a given type of feature, and whether the circle's interior is open or closed indicates presence or absence for a single attribute. This scheme can be effective on single-theme maps with straightforward titles. For example, a map of mine unionization might use filled-in cir-

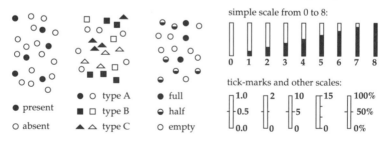

FIGURE 3.4. Point symbols can exploit a container metaphor to portray presence or absence (left) or a thermometer metaphor to portray a graduated linear scale (right).

cles to show unionized mines and open circles to show nonunion mines. A two-attribute map with filled and open circles, triangles and squares can employ different shapes to distinguish among three categories for one trait and a dichotomous filled-empty code to indicate the presence or absence of a second characteristic. Half-filled circles communicate intuitively an intermediate, "partly present" condition.

Filling a point symbol's interior can improve the reliability of quantitative linear scaling by adding a frame of reference that exploits a thermometer metaphor. Because the thin vertical rectangles on the right side of Figure 3.4 are analogous to vertical temperature gauges, the map viewer readily associates the darkened part of the rectangular column with the graduated scale in the map key. If the scale runs from 0 to 8, for example, the average viewer should be able to estimate values to the nearest unit.[7] The series of frame-rectangle symbols in the upper-right part of Figure 3.4 illustrate this precision. If the interior of the symbol is completely black, for instance, the value represented is obviously 8, whereas if only the lower half is darkened, the value is 4.

Adding a small tick-mark at the midpoint of the rectangular frame promotes even more precise decoding, especially for values near the midpoint of the range. The lower-right part of Figure 3.4 demonstrates tick-marks appropriate for various ranges of data values, as well as their corresponding labels for the map key. Two or more tick-marks might be useful for shorter scales; if data values range from 0 and 15, for example, two evenly spaced intermediate tick-marks would immediately identify the values 5 and 10. Because the vertical frame can be a standard for judging proportionality, the frame-rectangle symbol can portray not only counts or absolute magnitudes but also intensity measurements, such as a percentage scale ranging from 0 to 100. Indeed, the frame-rectangle symbol is a notable exception to the rule that size is a visual variable appropriate only for portraying differences in amount.

Line Symbols

In contrast to point symbols, cartographic line symbols have not only location but also the dimension of length. And as Figure 3.5 shows, lineal extent can exploit any of Bertin's six retinal variables. In addition to lateral and internal variations in size and value, line symbols composed of strings of tiny point symbols or dashes can vary in shape or texture. Because length affords a bigger stage for visual differentiation, medium-width line symbols can vary in hue more effectively than standard-size point symbols can. Linear features have intrinsic variations in shape and orientation; in addition, line symbols showing flows or one-way streets often require graphic marks to indicate direction.

As they are for cartographic point symbols, the retinal variations of line symbols are specialized. Shape and hue reflect differences among categories and types of features; size, value, and texture indicate differences in quantity or importance; and orientation shows direction by the use of arrows. Among the more common uses of line symbols, differences in thickness portray the relative magnitude of highway traffic, foreign trade, and other flows, whereas differences in texture show the relative importance of municipal, county, state, and national boundaries.

Although the size-magnitude relationship of flow maps is firmly grounded in graphic theory, use of texture variations to portray a hierarchy of political boundaries at least partly reflects cartographic tradition. Highway and street maps also have many traditional symbol-feature associations, such as blue lines to indicate rivers, streams, and canals. As part of a common graphic vocabulary, traditional map symbols usually promote efficient decoding.

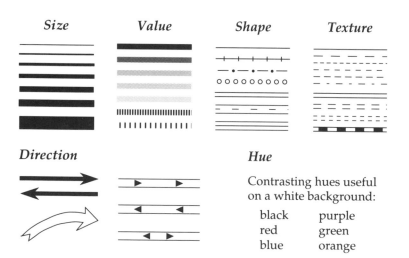

FIGURE 3.5. Common retinal variables for cartographic line symbols.

Not all traditional symbols are international in use, and what a map author may think is a standard symbol could easily confuse foreign readers conditioned to a different graphic idiom. A typical example is the thin solid line with short, evenly spaced, perpendicular cross-ticks used on North American maps to represent railways. Conditioned since childhood to associate this symbol with train tracks, most Americans can identify it instantly without reference to the map key. Yet this vaguely tracklike symbol is by no means obvious to European readers, accustomed to way-finding maps that represent railways with relatively thick, solid black lines. When addressing an international audience, the map author should be particularly careful to provide either examples in the key or unambiguous labels adjacent to individual symbols.

Traditional symbol-feature associations are sometimes so strong that violating them leads to misinterpretation or confusion.[8] For example, a map on which the traditional railroad symbol portrays a pipeline or political boundary will mislead or at least momentarily confuse many readers. Some semiotic codes are so familiar, in fact, that viewers might ignore the map key and decode the symbol erroneously. The map author should be both aware and wary of conventional symbols as used on road maps, topographic maps, and other widely circulated cartographic products.

Although conventions in symbolization usually reflect semiotic logic as well as cartographic tradition, some standard symbols are more natural or logical than others. For example, European map makers most likely rejected the North American cross-tick railway symbol as not fully coherent. Even though its evenly spaced cross-ticks are analogous to ties, or stringers, the overall symbol more logically suggests a monorail than the two parallel steel rails required for ordinary trains. Yet line symbols based on an abridgement or attenuation metaphor have broad international acceptance. As the examples in the left part of Figure 3.6 demonstrate, dashed or dotted lines are ideal for portraying highways under construction, unimproved dirt roads, abandoned railways, and the uncertain boundaries of culture or language regions. The most effective abridgement symbol is a chopped-up but otherwise identical version of the line symbol that represents the feature's complete or more fully functional counterpart.

Another form of attenuation metaphor is the visually recessive, "screened-back" layer of lines, labels, and other symbols
portraying less important features, provided as a geographic
frame of reference. As the right side of Figure 3.6 illustrates, uniformly gray line symbols form a background layer less likely to
attract the map viewer's attention than contrasting solid-black
symbols in the foreground layer. The readily decoded contrast
between layers of black and graytone symbols demonstrates the
effectiveness of value as a retinal variable for communicating relative importance.

Constraints on the thickness of line symbols restrict the flexibility and use of some retinal variables. Thin and medium-weight
lines, for instance, do not support a range of graytone values or
hues as effectively as comparatively thick lines, and very thin
gray lines sometimes disappear during printing. As in the right
part of Figure 3.6, line symbols must sometimes be thickened
before screening to prevent the graytone screen from obliterating
important details. These constraints can also be aesthetic; because very thick lines are visually cumbersome, especially if used
in great number, size has less flexibility as a retinal variable for
line symbols than for point symbols. Further constraints are im-
posed by the limitations of the drawing media, which might allow no more than six different thicknesses, and the human eye,
which can be unreliable in judging small differences in thickness.

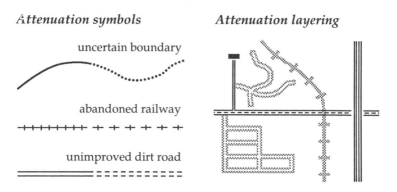

FIGURE 3.6. An attenuation metaphor suggests easily decoded line symbols
for uncertain, incomplete, or deteriorated linear features (left) and graytone
screening for differentiating background reference features from more
important foreground information (right).

Flow lines representing migration, banking transactions, telephone calls, and other transfers between pairs of points or areas can differ markedly from line symbols portraying linear features or boundaries. As the left part of Figure 3.7 illustrates, flow lines can jump across or circumvent intervening areas, cities, and other places to emphasize significant links between widely separated origins and destinations. By ignoring the transportation or communications networks that carry the traffic, leapfrogging flow lines use orientation to relate an origin to a destination. Even though theory might encourage scaling line symbols to make thickness proportional to size of the flow, the standard practice is to group the data values into no more than five or six magnitude categories represented by easily differentiated, progressively thicker flow lines.[9] Size or value may serve as a secondary retinal variable to portray the stream's relative importance, but direction has the dominant role, demanding prominent, unambiguous arrows. Flow symbols would hopelessly clutter a map that attempted to represent all possible flows between pairs; the map author might need not only to invoke a threshold to exclude minor flows, but also to use separate maps for major origins or destinations with multiple links.

Another atypical line symbol that employs orientation as a retinal variable is the *isoline*. As the right half of Figure 3.7 demonstrates, an isoline signifies a particular numerical value and tracks a continuous path between adjacent points sharing that value. But because the data points (shown as X's in Figure 3.7)

Flow linkages   *Isolines*

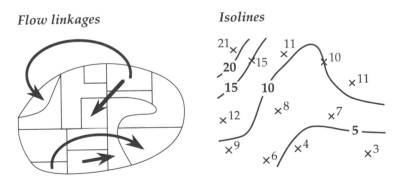

FIGURE 3.7. Orientation is an important retinal variable for flow lines (left) and isolines (right).

are sparse and usually reflect measurements not represented by their own isolines, the map author must estimate the paths of the isolines by interpolation, a process of educated guessing. Interpolation assumes that the measurements reflect a straightforward surface, as in the lower right-hand part of Figure 3.7, where the isoline representing a value of 5 passes halfway between data points with measured values of 3 and 7. When passing between 9 and 12, the isoline representing 10 logically lies closer to 9 than to 12. When the isolines have a constant step difference, or *contour interval*, their spacing reflects gradient or change. In the example, the contour interval is 5, and the gradient is greatest in the upper-left quarter of the map, where comparatively close lines reflect more marked local differences. Used commonly in the physical sciences to portray atmospheric temperature, barometric pressure, terrain elevation, and other environmental phenomena, isolines are sometimes used by social scientists to represent the geographic patterns of such diverse quantitative phenomena as urban real-estate values and the diffusion of an innovation.[10]

Area Symbols

Like point and line symbols, area symbols rely more naturally on some retinal variables than on others. As the examples in Figure 3.8 illustrate, the area symbol's two dimensions provide a generous stage for subtle variations in value, texture, shape, and hue. In contrast, the fixed boundaries of nonoverlapping areal mapping units obviate variation in size, and the one-dimensional nature of direction limits the role of orientation as a two-dimensional symbol. For maps showing magnitude or direction for areas, the most logical strategy is to place a graduated point symbol or arrow near the center of each area.[11]

The suitability of area symbols also depends upon the projected size of area features and their relative importance. Size is important, because area symbols describing large areas might distract the viewer from smaller, more important features, whereas area symbols describing tiny areas might be only marginally noticeable, as well as difficult to decode. For large and medium-size areas of secondary importance, a combination of interior labels and distinctive boundary symbols usually provides a suitable, visually recessive description. In contrast, area features too small to contain an easily identified, unambiguous patch of their area symbols might require point symbols, supplementary labels, or a detail inset map to make them more conspicuous and intelligible.

Area symbols can be ideal for maps showing geographic trends and patterns based on areal-unit data. Appropriate for such units as countries, provinces, states, counties, or census tracts, area symbols are also useful for mapping units formed by a land-use classification or other special partitioning of the region. By collectively filling the map, area symbols can focus the viewer's attention on the coherence or fragmentation of regional patterns. Potentially meaningful pattern elements revealed by area symbols include broad zones of relative uniformity, narrow bands of sharp contrast, and significant outliers or exceptions to an otherwise uniform or coherent pattern.

As shown in the left half of Figure 3.8, value and texture are useful retinal variables for a graduated series of progressively darker area symbols. These graduated series provide a logical, easily decoded, unambiguous representation of rank order for quantitative data grouped into categories. The graphic logic here is simple: the darker the symbol, the higher the value it represents. When a graphic series varies only in value or only in texture, the viewer can readily determine which of two differently coded areal units belongs to the higher category. The darkest area symbol indicates the spatial pattern of the highest data values, and the lightest area symbol indicates the pattern of the lowest data values.

Figure 3.8 also illustrates the efficacy of shape and hue as retinal variables for area symbols portraying qualitative variations. Because area symbols that differ in either shape or hue *look* dif-

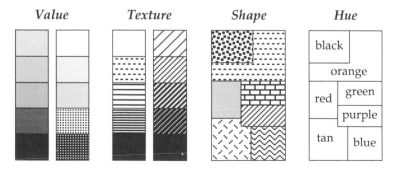

FIGURE 3.8. Common retinal variables for cartographic area symbols.

ferent, the viewer readily infers that the areas *are* different. Moreover, because variation in shape or hue does not suggest an ordered sequence, the map viewer is unlikely to misinterpret differences in kind as differences in value.

Devising a reliably reproducible, unambiguous series of progressively darker graduated area symbols for quantitative data can be challenging. Because fine-textured graytones tend to darken or bleed when printed, screen texture is important. As the two "Value" columns of Figure 3.8 illustrate, dot-screen area tints consist of rows or lines of tiny dots; the number of lines of dots per inch describes the symbol's screen texture. A screen with 65 lines per inch is coarser than a 150-line screen, which will have many more rows of much smaller dots. With high-quality presswork, comparatively fine screen tints, for example, 100-, 133-, or 150-line dot symbols, yield aesthetically pleasing graytones. For newspaper work and low-quality printing in general, using comparatively coarse 55- or 65-line screens is a prudent defense against ink spread, which can cause the tiny dots of a medium-gray area tint to grow and coalesce into a blotchy dark gray. A safe strategy for maps sent to professional journals or cost-conscious book publishers is to sacrifice aesthetics for graphic stability by using somewhat coarse, printer-proof graytones. But be wary about the final reproduced size of the map: if a map 6 inches wide is to be reduced to fit a column of type 3 inches wide, the map author must plan for a 65-line screen, for instance, by using a 32.5-line screen for the original, "up-size" artwork.

Three other principles can guide the cautious cartographer who decides to use graytone area symbols for quantitative data. First, use a uniform screen texture to avoid an embarrassing reversal of the sequence of graytones. Because ink spread attacks fine graytones more viciously than coarse graytones, a graduated series with a comparatively fine-grained symbol in the middle is particularly vulnerable to a gray-scale reversal in which the graytone for a middle category might appear darker in print than the graytone for the next higher category. Second, don't use more than five graytones between white and solid black. And third, take advantage of greater visual differentiation at the lighter end of the gray scale. Visual contrast is higher, for example, between area symbols that are 5 percent and 15 percent black than between symbols that are 45 and 55 percent black.[12]

Some cartographers prefer the greater graphic stability of parallel-line and crossed-line area symbols. As demonstrated in the "Texture" columns of Figure 3.8, these markedly coarser symbols mimic variations in graytone by varying the spacing and thickness of the lines that make up the pattern, by using dashed as well as continuous lines, or by adding a second set of perpendicular lines. No more than a quarter of the interior area of a texture-coded symbol typically is covered with ink or toner, so one or more texture-based rules usually supplements the modest variation in darkness. In the second "Texture" column of Figure 3.8, for example, the two lighter symbols associate more closely spaced lines with a higher value, whereas the middle three symbols relate thicker lines to higher values. The resulting series is coherent and easily decoded, because its two texture rules not only reinforce the darker-greater metaphor but don't interfere with each other. The graduated series in first "Texture" column also works, because its texture rules neither overlap nor conflict with the darker-greater metaphor: the shift from dashed to continuous lines occurs between the second and third lightest symbols, and the spacing rule affects only the third and fourth lightest symbols. Employing several texture rules in the same series can be a useful strategy for increasing the number of steps in a graduated series of area symbols.

Using different texture rules *in the same part* of a graduated series invites confusion and misinterpretation. The four graphically inconsistent series on left side of Figure 3.9 would make map reading tedious by forcing the conscientious viewer to refer repeatedly to the map key. For instance, the single-direction line patterns at both ends of the third series create a logical conflict between the darkness rule and the number-of-directions rule. While the careful map reader must pause to consider whether a dark single-line pattern can represent a higher category than a light crossed-line pattern, the careless reader who ignores the map key might assume erroneously that the dark single-line symbol at the bottom of the column falls between the lighter single-line pattern and the lightest of the three crossed-line patterns. After all, it seems only logical that crossed-line symbols represent greater data values than single-line symbols. In contrast, the logically consistent graduated series on right side of Figure 3.9 are efficient, effective, reliable, and quickly decoded. A continuous and noticeable increase in darkness is everywhere consistent with a reduction in line spacing, a shift from dashed to continuous lines, or a shift from from single lines to crossed lines. When a texture rule reinforces the darkness rule, the map viewer easily comprehends the graphic logic needed for accurate decoding.

Despite their graphic stability, relatively coarse line-pattern area symbols lack the eye appeal of well-planned, carefully reproduced fine-dot area symbols. When the need for reliable decoding dictates the use of comparatively harsh line patterns, three strategies can be employed to reduce the assault on the map viewer's sense of aesthetics. First, avoid vertical single-line patterns, which seem to vibrate; single-line patterns diddle the eye least when the lines are horizontal. Second, make full use of crossed-line patterns, which are less "busy" than single-line patterns; diagonally oriented cross-line patterns usually look better than criss-crossed webs of horizontal and vertical lines. Third, make texture less blatant by minimizing the separation of adjacent parallel lines in the lightest, least dense line pattern. The fifth and sixth columns in Figure 3.9 illustrate effectively that the need for coherent texture codes can encourage the use of coarse, visually obnoxious symbols, which would especially be a problem here if a large part of a map had to be coded in one of the patterns at the top.

Coarse-textured symbols are also unsuitable for small areal

Ambiguous series   Straightforward series

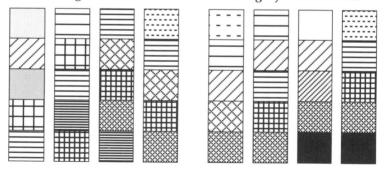

FIGURE 3.9. (Left) Ambiguity in a graded series of area symbols because of conflict among graphic rules relating higher values to (1) decreased spacing, (2) shifts from parallel-line to crossed-line patterns, (3) shifts from dashed to solid lines, and (4) darker overall tones. (Right) Coherence of texture rules and the darkness rule yields a straightforward, progressively darker series of area symbols.

units with low data values, which might retard or stymie decoding by showing only one parallel line—or worse, none. If the map also employs clear white as an area symbol, every instance of the lightest line or area pattern should provide the unambiguous spacing clue of at least two parallel lines.

Aesthetics and efficient decoding are also important for maps showing different kinds of areas. In choosing qualitative area symbols, the map author must balance the need for visual contrast with a disdain for ugliness. The principal visual variables for portraying differences in kind are shape and color (illustrated by the two rightmost columns in Figure 3.8). Since the cost of reproduction precludes the use of color for most publications in the humanities and social sciences, the map author almost always must resort to area symbols that differ in shape.[13] If electronic graphics are used, the range of choices includes not only dot, single-line, and crossed-line patterns but also a variety of "pattern fills," some of which vibrate, draw the eye, or suggest rows of ball bearings or the cane-woven seats of antique chairs. These area patterns may look different and even intriguing on the graphics system's menu of fill patterns. Some even have a grain of semiotic logic; for instance, wavy horizontal lines might suggest a body of water. But a discordant jumble of these graphic clichés plastered across a map suggests that the author is a hacker with a new toy, rather than a scholar attuned to geographic relationships.

A few general principles can guide the selection of qualitative area symbols. First, use similar patterns for similar categories and relatively dissimilar patterns for more distinctly different categories. On a map of land use, for instance, symbols showing lower- and higher-density residential land use might both be diagonal cross-line patterns, with the lighter, coarser of these two symbols representing the lower-density land use. An irregular dot pattern for commercial land and a solid black area symbol for industrial land would provide a more marked contrast in texture, to emphasize the more substantial functional differences among residential, commercial, and industrial land-use categories.

Second, exploit with care such semiotic relationships as bushy-looking patterns for forest land and wavy patterns for water. Some wavy area symbols vibrate obnoxiously, and the metaphor seems inappropriate for small lakes and ponds. Easily decoded feature labels (such as "Walden Pond") often allow more subtle, visually harmonious contrasts among area symbols.

Third, consistency in the assignment of specific symbols to specific feature categories is useful in a series of maps. For example, in a history of World War I, variations from map to map in the symbol representing German-held territory would confuse the reader. In comparison, a reader can use a familiar graphic vocabulary learned from the first few maps in a book or article to unlock later maps efficiently and reliably.

Two area symbols with special aesthetic and practical limitations are solid black and clear white. Because black draws the eye and monopolizes the viewer's attention, the size of the average feature in the category portrayed with black and the category's areal extensiveness are both important. For example, on a land-use map concerned primarily with the shape and distribution of villages, use of solid black to represent farmland would emphasize a less significant category and produce a dark, dismal image of the region. For maps of qualitative area data, reserve black for very important and comparatively less common features. In contrast, white is a visually recessive area symbol useful for portraying water, missing data, or other areas not part of the map's typology of important features. Because of the "nothingness" metaphor, based on the absence of ink or toner, white usually is a poor choice for an important areal feature.

FORM AND FUNCTION IN CARTOGRAPHIC REPRESENTATION

As elements in a constructed view of reality, cartographic symbols must communicate differences among geographic features, functional relationships, and relative significance. This section examines for a few typical yet diverse cartographic applications how the map author can orchestrate Bertin's retinal variables to promote the theme or goal of the map. The examples that follow are generally typical graphic models selected to address many, if not most, situations in which humanists and social scientists might want to use a map. In addition to integrating point, line, and area symbols, this section explores competition among symbols for space on the map and for the attention of the map viewer.

Portraying a Location and Its Neighbors

The most common map in expository cartography is the simple locator map that describes a place or feature in relation to nearby places or features with which the viewer might be acquainted.

When writing about journalistic cartography, I often subconsciously precede the term "locator map" with the adjective "simple," as I have done here.[14] But a simple map of commonplace features can be both essential and elegant, and even the simple map can be done poorly and fail. Its design reflects a variety of decisions—scale, projection, how large an area to show, which features to include, whether to use an inset. I ignore such decisions in this chapter in order to focus on the role of retinal variables.

Figure 3.10 describes the location of the Lorenzo Historic Site, in Cazenovia, New York. (Lorenzo was built in 1807 and was the home of Joseph Ellicott, chief surveyor and resident land agent for the Holland Land Company, a huge private land-selling enterprise in central and western New York.) Designed to explain a specific location that the reader might want to visit, this map is typical of locator maps for studies in architectural history, historical preservation, and industrial archaeology and reflects a careful selection of present-day features. Inclusion of portions of major transportation links and one or two large, better known places provides a geographic frame of reference for a smaller,

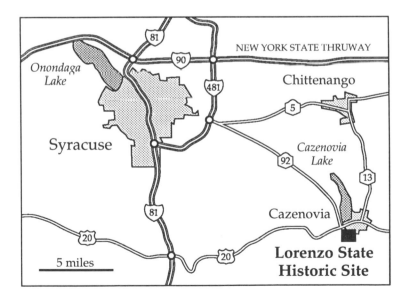

FIGURE 3.10. Locator map for the Lorenzo State Historic Site, in Cazenovia, New York.

more obscure place. Because the focus of the map is on transportation routes, point and area symbols are much less numerous than line symbols. A prominent black square representing Lorenzo, the focal point of the map, differs in shape from the open circles representing expressway interchanges. Area symbols with straightforward labels have subtle differences in the shape of internal elements to distinguish urban places from lakes; because of the map's labels and context, a key is unnecessary. Shape also distinguishes expressways such as the New York State Thruway and Interstate 81 from common highways, and the familiar shapes of highway shields help the viewer distinguish among route numbers for the Interstate, U.S., and New York road systems.

The use of the qualitative retinal variables shape and texture in this example can be readily generalized to other situations. A smaller-scale map concerned more with Lorenzo's position within North America might use lines varying in texture to distinguish state and international boundaries. Shape would differentiate these boundaries from rivers and shorelines, and dots varying in size could distinguish large cities from small ones. A locator map showing Lorenzo in the nineteenth century might use line symbols differing in shape to show roads, canals, and wagon roads, and area symbols varying in shape and texture to identify the Holland Purchase, the Military Tract, the Indian reservations, and other significant land holdings in the region. A large-scale map describing an archaeological dig might be deliberately vague about transportation routes, yet use several area symbols differing in shape to represent water, bare rock, marshland, and woodland.

Portraying Causal Relationships among Features

Adding causally relevant features can convert a map showing only location to a map offering an explanation or interpretation. Careful selection of causally related factors can provide not only a concise description of where a feature is, but also a cogent argument about why it is where it is, or perhaps of greater interest, why it isn't where it isn't. For example, a simple locator map of Chicago tells the viewer very little about why an enormously important city developed at the southwestern end of Lake Michigan rather than elsewhere in the region. A logical geographic explanation for Chicago's development lies hidden in standard reference maps of the United States. To make this explanation apparent, the map author must focus on the late-nineteenth century, show the convergence at Chicago of numerous railroad lines collecting agricultural products from a large part of the Midwest, and indicate the low-cost shipping route through Lakes Michigan, Huron, and Erie to Buffalo and other milling centers and railheads in the East. The retinal variables size and shape would be particularly useful in shifting the viewer's attention from features appearing merely as part of the map's geographic frame of reference to features essential to the map author's explanation or causal interpretation.

FIGURE 3.11. Effects of a southwest wind, Lake Michigan, and the North Branch of the Chicago River on the area destroyed by the Great Chicago Fire of 1871.

Figure 3.11 uses the great Chicago fire of 1871 to illustrate the juxtaposition of causally connected features. A dark, prominently labeled point symbol marks the location of O'Leary's barn, in the neighborhood of wooden cottages and outbuildings where the fire started shortly after 9 p.m. on Sunday, October 8, 1871.[15] The fire quickly got out of control, and a brisk southwest wind pushed the blaze through the downtown business district, which is roughly indicated by four point symbols representing the city's major railroad passenger stations. The map shows the burnt-over area, bounded by the shore of Lake Michigan on the east, the north branch of the Chicago River on the west, and Lincoln Park on the north. That the area south of O'Leary's barn was spared attests to the wind's dominant role in spreading the fire. The 1871 city boundary provides a base for judging the relative extent of the holocaust, and the sparse network of selected major streets presents a fuller geographic frame of reference for readers familiar with the city.

Figure 3.11 also demonstrates how a map's line symbols can provide both contrast and harmony. Lines representing unlike features rely on marked differences in size (boundary of the fire zone), shape (streets), texture (city boundary), and graytone value (rivers). A heavy closed boundary line delimiting the burned area unites with the point symbol for O'Leary's barn to focus attention on the relationship between the extent and origin of the fire. An eye-catching solid black area symbol for the burnt-over area would have visually overwhelmed the barn, presented the extent-origin relationship less effectively, and eliminated symbols showing the railroad stations, park, river, and streets within the fire zone. If there were no lake or river, a medium-gray area symbol for the burned area might have offered an equally effective yet more aesthetically pleasing solution. Because the essential features on any complex map compete for a limited number of visually distinct symbols, the nature of the causal relationship and the unique geographic configuration of important elements challenge the map author to find a graphically effective compromise.

Portraying Routes

Maps can be vastly superior to words alone in describing routes. Route maps can help humanists and social scientists avoid deadly dull verbal descriptions of itineraries and networks and can present the reader with logically organized, readily accessible information. The varied uses of route maps include describing the travel experiences of an important writer or philosopher, portraying the wanderings of an explorer and the advances and retreats of an army, and illustrating the geographic structure and orientation of commodity movements and information flows. Retinal variables important on route maps include orientation to show direction of movement, size (thickness) to show relative importance, and shape and graytone value to provide contrast among background features and routes of different types.

Figure 3.12 illustrates how a route map can not only describe a complicated itinerary but also add dramatic visual emphasis. An economist or policy analyst might use this map to demonstrate the continued need for marine freight rail service across the lower Hudson River between New Jersey and Long Island. Without car float and lighterage service in New York harbor, rail freight from areas south and southwest of the region would be routed up the west side of the Hudson to Selkirk, then across the

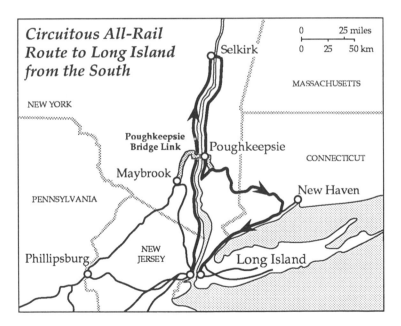

FIGURE 3.12. The extravagantly devious all-rail freight route from New Jersey to Long Island is a strong argument for maintaining car float and lighterage services across New York harbor.

river and down the east side to a junction about twenty miles south of Poughkeepsie, then further east along a meandering route almost as far as New Haven, and finally southwest along the shore route to a bridge crossing onto Long Island. Calling this path "circuitous" is far less effective than mapping it out. The map further simplifies an otherwise complex discussion by showing how an alternative "high-level" route across the Pough-keepsie Bridge might cut off an awkward diversion north to Sel-kirk, yet not escape an eastward diversion toward New Haven. The economist or economic historian might include this inevita-ble eastward detour into Connecticut as one reason for Conrail's decision to abandon the Maybrook–Poughkeepsie link after a fire severely damaged the bridge in 1974.

Size, shape, and value all have important roles in differentiat-ing the various linear features in Figure 3.12. A thick solid black line marks the circuitous all-rail route from New Jersey to Long Island as the map's dominant feature. This devious Selkirk route contrasts strikingly with the wider but medium-gray line symbol for the route across the Poughkeepsie Bridge, and with the thin-ner lines representing other relevant railroad lines from the west and south and on Long Island. These connecting rail lines are foreground features, which stand out from the less important light-gray state boundaries in the background. Differences in graytone also provide a clear contrast between land and water. The distinctive shapes of circular point symbols signify impor-tant junctions. Because of concisely worded labels and logical retinal variables, the map requires no key to explain its symbols.

Portraying a Phenomenon with Multiple Instances

Unlike the well-defined pattern of the area burned in the Chicago fire in Figure 3.11, many geographic phenomena consist of a multitude of separate, seemingly isolated occurrences. Indeed, the scholar might not know that a pattern exists at all until these occurrences are mapped, however crudely. In some cases the map confirms or rejects the humanist's hunch or the social scientist's hypothesis, in others the map as a two-dimensional inventory sheet serendipitously reveals a blatantly significant or intriguing-ly enigmatic relationship. Even the absence of a noteworthy spa-tial pattern can at times be meaningful. Maps are useful at both the discovery and confirmatory stages of research, and they may also help the author share these findings with readers.

Figure 3.13 illustrates the value of mapping multiple occurrences, in an example from onomastics, the study of names. Individual dots represent American counties or minor civil divisions with the name "Jefferson," presumably in honor of Thomas Jefferson.[16] Note the thin band stretching westward though Pennsylvania into the Midwest, the marked concentration of Jeffersons in Ohio, Indiana, Iowa, and Missouri, and the surprising rarity in the South, including Thomas Jefferson's native Virginia. This map is redesigned from one of hundreds of maps of place names, first names, surnames, ethnic restaurants, and other revealing surrogates used by historical geographer Wilbur Zelinsky to uncover significant regional differences in American culture.[17] Zelinsky mapped the distributions of other "blatantly nationalist" place names and found generally similar patterns for places named "Washington," "Franklin," "Union," "Jackson," and "Lincoln." Differences among patterns were minor; for example, Jacksons were only slightly more common in the South, which had about as many Lincolns as Jeffersons. This toponymic archaeology reflects the movement westward of settlers from the Pennsylvania-Delaware Valley "culture hearth." State laws regulating the naming and renaming of places might explain the puzzling anomaly in Illinois, and the reduced popularity elsewhere of nationalistic place names seems to reflect a preference for classical and British names in New York State and New England and reservations about the Republic, as well as less enthusiasm for patriot-heroes, in the South. As artifacts, regional patterns of place names, personal names, house types, and other aspects of the human landscape point consistently to the existence and influence of three psychically different American culture hearths, in New England, Pennsylvania, and the South.

Like most dot maps produced by Zelinsky and other cartographic archaeologists, Figure 3.13 shows just a single distribution. Its primary retinal variable is shape, which distinguishes the dots from the thin, visually recessive state boundaries in the background. Portraying several point phenomena simultaneously on the same map is false economy when the features are numerous and spatially related. A single map with round dots for Jeffersons, plus-signs for Franklins, and open squares for Washingtons, for instance, would have to be much larger in size; otherwise, the symbols would either overlap or require a magnifying lens for accurate decoding. Although it would reveal relatively empty areas in the South and West, such a complex and cluttered map would thwart efforts to detect more subtle differences among the three layers of information. Even if different primary colors were used, visually separating the three layers of foreground information would be tedious, if not impossible. For the very small point symbols used on maps to describe the overall pattern of numerous point features of different types, the retinal variables shape and hue are visually feeble and almost always ineffective.

As I have redesigned it, Figure 3.13 uses a larger dot than Zelinsky's original version, which was about twice as large in scale. Although my comparatively prominent dots more clearly raise the pattern of Jeffersons above the network of state boundaries, dots any larger than these would begin to look harsh and ugly. If places named "Jefferson" were much more numerous or densely clustered, I would need a smaller dot (or perhaps a larger map) to prevent coalescence. In general, dots should just barely merge in the most dense area because, in principle at least, individual dots

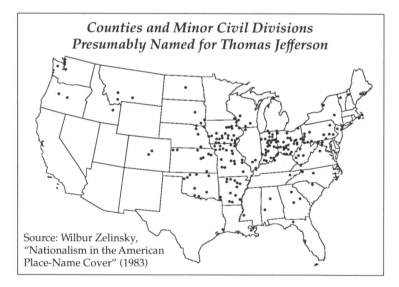

FIGURE 3.13. Map of counties, towns, and cities with the name "Jefferson" reveals a modest band through the Middle Atlantic region, widened to a stronger, broader concentration in the Midwest. Equally noteworthy is the rarity of places named "Jefferson" in New England and the South.

should be countable. The series of dot maps at a uniform scale, with a consistent dot size determined by the distribution with the greatest regional concentration, also illustrates in Zelinsky's article that places named "Washington" are more common than places named "Lincoln."

Portraying Variation in Count or Magnitude

In contrast to data that locate individual occurrences, count data reported by area or site require a point symbol that varies either in its overall size or in the numerousness of its internal elements. Figure 3.3 demonstrated how graduated point symbols scaled according to length, area, or volume use size as a retinal variable to portray magnitude. Because maps with graduated circles or other proportional point symbols are common, I chose for this section's example a less prevalent, yet sometimes highly effective, point symbol based on numerousness. The pattern of point symbols in Figure 3.14 would serve well a political scientist, economist, or economic historian who wants to show the marked regional bias in the pattern of savings and loan failures resulting from the deregulation of the American financial services industry during the 1980s. Because the dots are countable, the map requires no key. Size is a secondary retinal variable, but there is little likelihood that the viewer will underestimate the magnitude of the map's larger symbols.

Figure 3.14 uses the perceptual force of medium-size dots arranged in uniformly spaced rows and columns to form unified point symbols that stand out from the network of state boundaries in the background. Even where a few dots spill over into adjoining areas, as for Illinois, the structural regularity of the symbol's grid and squarish outline merges its constituent dots into a visually coherent whole. This regularity indicates clearly that the individual dots do not represent individual locations, as did the dots in Figure 3.13. Aligning the symbols to match a state's north–south or east–west elongation minimizes spillover and promotes a more effective association between symbols and areal units. Grouping the dots into rows or columns of five helps a viewer to retrieve the count for a particular state.

Numerousness doesn't always work.[18] Much of Figure 3.14's visual effectiveness and rhetorical impact arises from the absence or rarity of bank failures in many states, which must contribute nonetheless to the federally financed bailout. A phenomenon five or ten times as prevalent would require a key, perhaps with a few examples as well as a statement indicating the number of occurrences represented by a single dot. The geography of the savings and loan crisis also helps make the map's graphic coding scheme work; a phenomenon concentrated in the comparatively small states of the Northeast, for example, would require either interrupting the national pattern by a detail inset map or placing numerous symbols outside the state boundaries, as for New Jersey and Maryland in Figure 3.14. Most types of count data probably require a more conventional symbol, such as the vertical bar or graduated circle, but the map author should at least consider the straightforward simplicity and readily communicated graphic logic of a point symbol with countable elements.

Portraying Variation in Density

Were the pattern in Figure 3.14 not so heavily concentrated in a few states, the perceptive reader would surely ask whether the distribution shown merely reflects the existence of a markedly greater number of savings and loan associations, both solvent

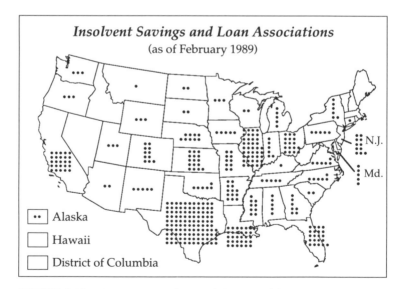

FIGURE 3.14. Numerousness of internal elements of these point symbols shows the marked concentration in a few states of the savings and loan scandal of the late 1980s.

and insolvent, in Texas, Illinois, Louisiana, and a few other states. To address this question, and perhaps to dramatize further the regional disparities of the savings and loan bailout, we might compute and map the ratio of each state's insolvencies to its total number of savings and loan associations. Deriving a ratio variable from a count variable and mapping both is a safe strategy for analyzing magnitude data. In many cases, though, all the author and the reader might really need is a map of the ratio.

Figure 3.15, derived from a map James Cerny included with an article published in *James Joyce Quarterly* under the title "Joyce's Mental Map," demonstrates the use of ratio maps in the humanities and social sciences.[19] Cerny was curious about the numerous river references in Joyce's enigmatic *Finnegan's Wake*

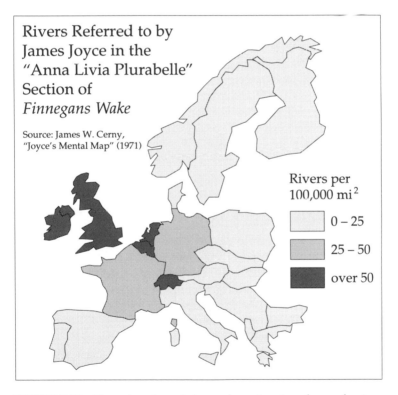

Rivers Referred to by James Joyce in the "Anna Livia Plurabelle" Section of *Finnegans Wake*

Source: James W. Cerny, "Joyce's Mental Map" (1971)

Rivers per 100,000 mi^2

☐ 0 – 25

☐ 25 – 50

■ over 50

FIGURE 3.15. Three-class choropleth map shows Joyce's preference for river names from countries where he had lived, nearby countries, and countries where English is the dominant language.

and what they might reveal about the author's knowledge of Europe. Other literary scholars had noted Joyce's use of more than eight hundred river references, such as "irrawaddyng" and "rhunerhinestones," which stem respectively from "Irrawaddy" and "Rhine." Apparently Joyce actively sought river names from peers and associates and wove them into his writing. Cerny examined various lists of river references, assigned rivers to countries or whole continents, and computed a ratio by dividing each country's river count by its land area. As Cerny's resulting three-category map of the density of river references indicates, Joyce's mental map reflects the areal extent of spoken English, periods of residence in Ireland, France, and Switzerland, and an inverse relationship with distance.

The map succeeds because of the logical link between density of river references and graytone value. Density is an intensity measurement appropriately portrayed by area symbols that vary in intensity. A single glance at the map key is all the viewer needs to associate the light-gray symbol with the lowest densities, the medium-gray symbol with intermediate densities, and the dark-gray symbol with the highest densities. Scholars working with quantitative data tabulated by area will find graytone value the most functionally obvious and readily decoded retinal variable for mapping percentages, medians, rates of change, densities, ratios, and other indexes that compensate for differences in size among areal units.

ACCESSIBLE CODING AND CARTOGRAPHIC GOALS

This chapter has presented a variety of strategies for cartographic communication and a number of rules for selecting functional and visually pleasing solutions. The discussion has underscored the need to promote ready and reliable decoding, to avoid conflict among coding rules and metaphors, and to ensure the accurate reproduction of an aesthetically acceptable printed map. I hope my examples have demonstrated the utility of Bertin's theory, as well as the need to exploit, or at least be aware of, graphic metaphors and cartographic tradition. Map authors who ignore the logic of retinal variables or of widely used symbol-feature associations risk confusing or misleading their readers. I hope, too, that this chapter's examples of maps that make otherwise complex information accessible to readers will encourage schol-ars to use maps wherever maps are needed.

Conscious and deliberate assessment of an expository map's goals is essential. To employ cartographic theory effectively, the author must decide which features or relationships to emphasize and whether to focus on differences in kind, differences in amount, or differences in intensity. In working out the goals of a particular map, the humanist or social scientist must examine both the promise and the limitations of the data and the map's supporting role as an illustration: Can the information available reveal reliably what we want it to show? How will our written text introduce, describe, and interpret the map? The next chapter looks more closely at the integration of map and text by explor-ing the functions of map titles, feature labels, keys, and source notes.

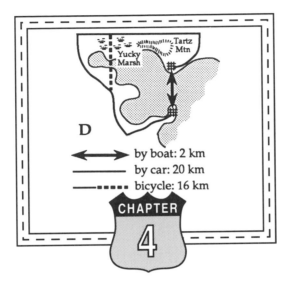

by boat: 2 km
by car: 20 km
bicycle: 16 km

CHAPTER
4

Map Goals,
Map Titles,
and Creative
Labeling

THE WORDS ON A MAP PROVIDE A NEEDED LINK BE-
tween the cartographic symbols and the natural language of authors and readers. This link is essential, because graphic symbols
employing the graphic metaphors and retinal variables discussed
in the last chapter are insufficient to describe the nature and attributes of many features or observations represented on maps.
Words that reflect the author's ideas and terminology are needed
to tie the map to the written text and to integrate the structurally
diverse realms of cartographic and literary representation. The
map key helps the viewer unlock the meaning of cartographic
symbols. Various labels identify particular places and features,
describe sources or warn about uncertainty in the data, and announce the map's title and intent. Moreover, a carefully worded
title or caption can explain why the author used the map and call
attention to a specific pattern or relationship.

This chapter begins by looking at the goals an author can set
for a map and how these goals can lead to a coherent map theme
and a meaningful map title. Defining and refining goals are essential to map authorship and require close and conscious coordination between writing the initial draft of the text and rough-
ing out a preliminary version of the map. The use of type as a
cartographic symbol is also examined in this chapter. Type distinguishes among different classes of features and indicates their
relative importance; it also provides a compact, readily interpreted alphabetic or numeric code, akin in many ways to Bertin's retinal variables.

COMMUNICATION GOALS, MAP CONTENT, AND GRAPHIC HIERARCHIES

Maps can communicate a well-defined message, serve as a conveniently placed geographic appendix for the interested reader, or
merely decorate pages and add variety to their layout.[1] Although
these functions may coexist, the social scientist or humanist best
serves the reader by identifying a specific message or goal for

each map. As an illustration to support the text, the map should complement the author's written discourse, not compete with it. And conscious recognition of what a map is to tell the reader greatly simplifies decisions about what features to include and which ones to emphasize.

References to Illustrations in the Text

Expository cartography works best when the author tells the reader to look at the map. After all, can the reader be expected to take seriously an illustration introduced by an offhand parenthetical reference, tacked onto the end of a paragraph or sentence? To write "The fire spread eastward as far as Lake Michigan (Fig. 17)" subtly invites the reader to ignore the map and move on to the next sentence. If the map is important to the text, say so—and if not, don't use it.

A short imperative sentence can direct attention to the map, as in "Look at Figure 17," or its less strident version, "See Figure 17." If this approach seems too authoritarian or stylistically harsh, consider encasing the command in parentheses. Having forcefully pointed out the map, the author should then state what the map shows and why this pattern or relationship is important.

Making the illustration the subject of a straightforward declarative sentence catches the reader's attention and justifies the interruption, as for example, "Figure 17 shows how a southwest wind spread the fire and how Lake Michigan and Lincoln Park helped contain it." Subsequent sentences might point out other important features on the map and guide the reader to a complete and accurate interpretation; for example, "Note that the burned area is east of the north branch of the Chicago River." A map that shows all places and features mentioned in the corresponding part of the text avoids confusing the reader.

Early in my career I encountered two journal editors who objected vehemently to such direct references to supporting exhibits. An illustration or table, they asserted, is an inanimate object and thus incapable of showing or possessing anything. Although they would accept a prepositional reference such as ". . . the relationship between . . . in Figure 17," under no circumstances would they allow a map or other display to serve as the subject of a sentence in the active voice. One editor even incorporated this dictum into his instructions to authors. As an untenured assistant professor, I did not argue further with these paragons of literary purity; their positions in the academic power structure assured that my published article reflected their preference, not mine.

A decade later, still aggravated by this wrongheaded attempt to separate words and pictures sharing a common goal, I wrote a short essay arguing that the direct reference to illustrations is not only logical but preferable.[2] One reference cited to support the acceptability of "Figure 17 shows . . ." was the enduring and authoritative *Chicago Manual of Style*, which supportively asserts "If there are many illustrations, they should be numbered, and text references to them should be by the numbers: 'figure 1 shows . . . ', 'see figure 2', '(fig. 3)'."[3] Although not as strong as I would prefer, this statement at least places the two direct references to the illustration ahead of the merely parenthetical one in the list of three acceptable options.

Convenient Layout and Tightly Focused Maps

Equally important in getting the reader to look at the map is a convenient page layout with exhibits positioned near the associated text. After all, the eyes move more rapidly than the fingers, and readers dislike having to hunt for inconveniently placed illustrations. Ideally, the map should appear above or below its text reference or on the facing page. But the map author usually has little control over the placement of illustrations, which the book designer or layout artist positions to promote visual balance. Like many products, books and journals are often judged by their packaging, and no publisher is going to risk alienating buyers or critics with a layout that looks amateurish.

Nonetheless, an author can promote a layout convenient to the reader by addressing each map to a single, compact part of the text and by not introducing illustrations prematurely. Text references to illustrations should be used sparingly, because mentioning a map too soon might encourage its placement well ahead of where the reader needs to examine it most closely. An initial parenthetical reference that merely points out a map discussed more fully later in the text might tempt the designer to ignore an insertion point marked on the manuscript or proof. *The Chicago Manual of Style* describes the designer's objective: "An illustration should be placed as close as possible to the first text reference to it. The illustration may precede the reference only if it is on the same page as the reference; otherwise, it should follow."[4] A compromise placement between the initial casual mention of the illustration number and the more complete examination of the map ten or more pages later can be distracting as well as inconvenient. It is far better if the author avoids the potential problem by making a single direct text reference to the illustration, followed immediately by complete discussion of the map's most important pattern or features. If further along in the book or article there is a need to refer to the earlier discussion of the map, a less distracting parenthetical reference is more appropriate than a direct reference.

Designing a map tailored to precise goals articulated in one or two consecutive paragraphs is almost always easier than forcing a single map to accommodate diverse objectives. A multiple-map strategy is especially appropriate it the author must examine various portions of a large area in detail or discuss several relatively complex distributions that could not coexist comfortably on a small, page-width map.[5] As chapter 7 demonstrates, multiple maps are particularly appropriate for historical essays. A progressive series of cartographic snapshots lets the author describe a complex situation at critical instants or periods of time and develop a coherent interpretation of a complex battle, a war, or the evolution of a city or a nation-state. And as chapter 8 explains, a series of progressive overlays on a visually recessive common base can be a very efficient method for discussing causal relationships.

Addressing multiple maps to compact portions of the text need not inflate publication costs because of redundant artwork. In most cases the alternative to a single map is two maps, not six or seven. Although a single large map is occasionally more useful than several highly focused maps, production costs escalate when a book must include fold-out maps tipped into the binding or a folded sheet map in a pocket attached to the inside back cover. Tightly focused maps are usually smaller than complex composite maps and need not consume excessive space. Moreover, the author who discusses clear, carefully composed cartographic drawings is less likely to waste words.

In deciding whether to use one map or several maps, the author must assess the phenomenon's potential for flexibility in the sequential presentation of facts, relationships, and arguments. Does the subject demand a single chain of reasoning, or are there two or more acceptable approaches? What are the cartographic demands of these options? Would a region-by-region treatment be logically coherent or distracting? Can the discussion move smoothly from an overview, with a small-scale map, to a series of case studies supported by individual large-scale maps? How might a distinctly different small-scale map at the end of the chapter effectively recapitulate and summarize the preceding discussion? Is a temporal sequence appropriate? What will each map show—what will it tell the reader? What sequence of geographic relationships most effectively illustrates or documents the author's thesis? Can features on the maps be grouped into themes? What foreground features are essential? What background features are needed, and must they appear on all maps, most maps, or just a few maps? The variety and complexity of these questions demonstrate the need to think carefully about cartographic needs while developing the outline of a manuscript and composing the first rough draft.

Historians and other scholars often provide one or more general reference maps, commonly bound into the book at the front for ready access. As the antithesis of the focused, goal-oriented map I advocate throughout this book, the reference map is somewhat a special case. It needs to describe the region in which the important events occurred (such as exploration or settlement), and it needs to show the specific places, mentioned in a variety of places in the text, that the reader is likely to consult it about—a challenging task, to be sure. Although they are useful if conveniently placed, reference maps must address a broad and inherently vague goal that requires a careful appraisal of their content and symbolization. If used at all, reference maps should be supplemented by other maps with specific goals more easily related to specific passages in the author's discourse. Because the two types of maps can be complementary, the author should choose their content, labels, symbols, and projections to promote integration and coherence.

Telling Titles

Poor integration of maps and text often results from thoughtless use of staid traditional titles that hide the message and function of the map. The title commonly has the largest, most prominent type on the map, and to clutter it with obvious or marginally significant information is to waste a valuable opportunity to tell the reader what the map shows that is new and important. In a book about London, for instance, an author who presents several maps of London has no need to include the name of the city in the title of every map. If a citywide map is intended to portray the comparatively primitive development of the sewer system in poorer neighborhoods, its title should mention both sewers and poverty and should call attention to their inverse relationship. If the map is one of a series of closely positioned maps showing changes over time in the African-American proportion of the population in Boston's census tracts, the key new fact that the title should display prominently is the census year. The reader readily misses the point of a blandly titled map in which the most important words are buried in the key or a subtitle.

Map titles should display new or variable information more prominently than other facts that describe the map. Information held constant across several maps might be relegated to a subtitle or described only in the caption set in type below the illustration. Common information might include the instant or period of time, the place, or the name of the geographic distribution or phenomenon. Whenever any of these factors varies from map to map, the title should prominently inform the reader how each individual map is different.

Map authors can choose between divided and declarative strategies for composing map titles. The divided strategy separates the various parts of the title with line breaks or commas, as in "Registered Voters, Percentage Increase by Precinct, 1970–1980." In contrast, the declarative strategy uses a complete sentence, with a subject, verb, direct object, and usually one or more prepositional phrases, to tell the reader what the map shows, as in "During the 1970s African-American neighborhoods experienced the greatest increases in percentage of registered voters." Capitalizing only the first word and proper names encourages the viewer to read the title as a sentence. Although excellent for injecting meaning into the map and directing the reader's attention to important patterns, declarative titles perhaps too easily incorporate the author's biases or steer the reader away from other, potentially significant patterns or interpretations. Moreover, declarative titles readily incorporate causal explanations that the author should support with additional evidence; for example, "Because of registration drives sponsored by Democratic and religious groups, during the 1970s African-American neighborhoods experienced the greatest increases in percentage of registered voters." Nonetheless, a straightforward declarative title is the best descriptive heading for an uncomplicated map with a clear message.

Map titles call for precise, minimalist wording. In composing a divided title, the map author must not only be concise but also anticipate possible confusion. Context and variety of data often will indicate which aspects of a mapped distribution are essential elements of the title. For instance, in a political scientist's study relating registration and voting rates mapped by precinct to ethnicity mapped by census tract, the title "Registered Voters, Percentage Increase by Precinct, 1970–1980" not only identifies and highlights the map's theme but points out that the data are percentage rates of change computed for precincts, not census tracts. The need to avoid confusion calls for particular care in the ordering of multiple elements in a declarative title. Although "African-American neighborhoods experienced the greatest increases in percentage of registered voters during the 1970s" might place the focus of the author's interpretation in the lead position at the beginning of the sentence, the version "During the 1970s African-American neighborhoods experienced the greatest increases in percentage of registered voters" more clearly links "greatest" to a comparison with other neighborhoods and thus prevents the reader from concluding merely that African-American registrations increased more during the 1970s than in other decades.

Brief titles sometimes require clarification. A subtitle below the main title, or a descriptive heading above the map key, can supplement or clarify a short title requiring a more precise definition of the measurements or features. For example, if the title "Urban Growth: 1980–90" is set in large type to announce the phenomenon mapped and the decade, the carefully worded subtitle "Percentage Rate of Increase of the Population Residing in an Urban Area, by County" indicates clearly that the map refers to percentage rates, not absolute numbers, and that "Urban" refers to the population of the urban areas within each county, not the land area measured for a category of land use.[6] Because this map would require a key, the subtitle might conveniently be placed over the key as a heading. Some maps require additional text on the map or in the caption, to identify the source of the information, provide an essential definition, or warn of uncertainty in the data.

Authors might choose a cartographic style that avoids prominent type on the map by using the caption as the principal title. I don't recommend this approach, because a conspicuous and concise title can be important in attracting the reader's attention and communicating the theme or message of the map. But even with this caption-only strategy, a subtle and concise heading directly above the map key can promote a clear, unambiguous interpretation of the symbols and the pattern or trend they represent.

Working up a cartographic style sheet can be useful for a publication with many maps. A style sheet can assist the author to use fonts and capitalization consistently and to avoid distractingly haphazard titles, subtitles, notes, and captions. For a publication with relatively few maps, the author usually needs to plan carefully only one typical map and to follow its style in designing the others.

The title and the key are important elements in a map's overall layout. The layout can effect the integration of the title and the key and the title's effectiveness. The traditional positions—the title at the top of the map and the key at the bottom—separate these elements and thwart their coordination. Figure 4.1, which shows two schematic layouts for a squarish map of a tall, somewhat narrow region, demonstrates that the traditional approach (on the left side) may waste space and may reduce the size, scale, and detail of the mapped area. Even in a more appropriate narrower frame, with less wasted space, the separation of the title and key would inhibit an easy link between the measurement or theme described in the title and the symbols described in the key.

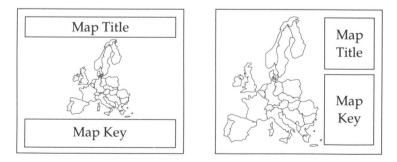

FIGURE 4.1. Traditional layout (left), with the title separated from the map key, contrasts with an alternative layout (right) that integrates the title and key and favors the prominent display of important parts of multiline titles.

In contrast, the nontraditional solution (on the right) not only permits a larger, more detailed map within the same frame but also favors a closer integration of the title and key. In this case, using a subtitle to serve simultaneously as the header for the key might eliminate some redundant words. Moreover, breaking the title into two or more short lines to fit the narrow space promotes the prominent display in large type of an important component of the title, such as the year-date.

Figure 4.2 shows three other layouts that integrate the title and key. The leftmost example shows the map in Figure 4.1 in a vertically elongated frame, which could be positioned on the page to the left of its caption if the designer would permit. The example in the center is of multiple maps sharing the same title and key, which show a region in different years. The example on the right shows the title and key occupying a generous portion of the frame left vacant by an irregularly shaped area. Rather than deciding automatically upon a traditional title-at-the-top layout, the map author should experiment with a variety of formats and evaluate their ability to provide the necessary detail within the space available.

Careful consideration of the tradeoff between the complexity of the map's symbols and the length of its title can sometimes lead to simplifying the map by making it possible to eliminate the key entirely. If the author needs only to point out those parts of a city where more than half of all households are poor, titling the map "Census Tracts with More Than 50 Percent of Households

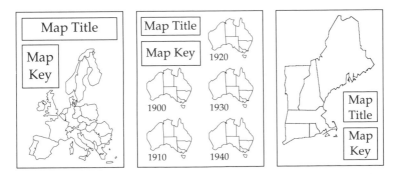

FIGURE 4.2. Three examples of layouts that promote the integration of the map's title and key.

below the Poverty Level" and highlighting those tracts with a single light graytone area symbol provides a simple, effective solution. If the 50-percent poverty level is indeed a reliable threshold for identifying poor neighborhoods, this simple two-class map is far more direct than a four-, five-, or six-category map with the slightly shorter title "Percentage of Households below the Poverty Level." The two-category map not only avoids a needlessly complex set of area symbols, but also allows room for map labels to show other useful features, such as specific neighborhoods, principal streets, or important landmarks. Although it contains less information about the geographic pattern of urban poverty than a multicategory map, the simpler map may more effectively support the author's discussion.

Scale of Concepts

A "scale of concepts" can guide the selection of a map's content, as well as its title, labels, symbols, and key. Simply stated, the map author needs to identify the features and relationships to be shown and their relative importance. Although these specifications might take the form of a written list, a simple sketch will do. This sketch should establish priorities by forcing the author to choose which features to emphasize or link together, which ones to include only recessively, and which ones to suppress altogether. It should differentiate foreground information directly related to whatever the author is discussing from background information used to tie the discussion to the reader's knowledge of the area. In deciding on background information, the author should consider maps discussed earlier in the book or article. After all, features presented or reinforced on previous maps provide a more reliable model of the reader's geographic frame of reference than imprecise notions about "what any intelligent person might be expected to know."

In addition to tying the map's content to its associated text, the scale of concepts simplifies the selection of symbols and the phrasing of the title. A map works best when the graphic hierarchy of its symbols reflects the conceptual hierarchy of features and relationships and draws the reader's attention to the most important fact and patterns. Because design is a holistic process, a sketch is more appropriate than a list for evaluating alternatives and choosing an effective design. Although an experienced cartographer with a good intuitive sense of design might bypass a preliminary drawing, the concept sketch, or "rough," is usually useful for making certain that all the essential details have been worked into an elegantly efficient design.

Focusing a map on an important relationship, rather than on a distribution or a single location, can be difficult for an inexperienced map author guided largely by a vague sense of tradition and acceptability. Consider the challenge confronting an anthropologist or rural sociologist who wants to show that two towns on opposite sides of a small bay are 2 km apart by boat, 16 km apart by bicycle, and 20 km apart by car. Obviously the map must include the two places, the shoreline, the intervening body of water, and at least the principal roads linking the towns. But how overtly should the map compare the different routes, and with what level of detail? Clearly words alone can convey the basic differences in distance. But how much detail should the map provide about obstacles? Is it merely the bay that makes the overland route longer? How circuitous is the overland route? Why is the bicycle route shorter? To what hazards does the shorter of the overland routes expose the cyclist, and is that route likely to be impassible in wet weather? Is the shortcut permanent or seasonal? Does the marine route have any hazards or significant obstacles, such as swift currents, dangerous tides, or shoals? Should the map rely largely on a scale bar to describe distance relationships, should its title mention relative distances, or should the map directly label distances for various parts of the overland routes? Answers to the questions follow from careful consideration of what the reader needs to be told about these routes and the ease of travel and of how the written text and the map might supplement and reinforce each other.

Figure 4.3, which describes schematically a few approaches to selecting features and labels, demonstrates the trial-and-error graphic reasoning that concept sketches promote. Example A shows how the map author can use labels in so-called balloons to point out and describe the available routes. This approach highlights a straightforward comparison of distances. Example B provides more precise measurements for individual route segments and notes that the cyclist can take advantage of a shortcut across a railway bridge and through a swamp. Its title focuses attention on this shortcut, portrayed with a bold dashed line, but its labels and symbols omit direct suggestions of hazards. In contrast, Example C calls attention to tidal bores and treacherous shoals in the bay and to the need to walk the bicycle across a railroad bridge when using the shortcut. Example D mentions no hazards, notes the length of various alternative routes in its key, and presents a comparatively generalized description of their shape. Yet this map can usefully explain how the configuration of the bay and the presence of a large swamp and a long, linear ridge yield an overland route eight to ten times longer than the marine route. Labels naming these features suggest their importance and provide unambiguous links to the author's discussion.

The four examples in Figure 4.3 show how experimenting with simple concept sketches fosters the integration of map and

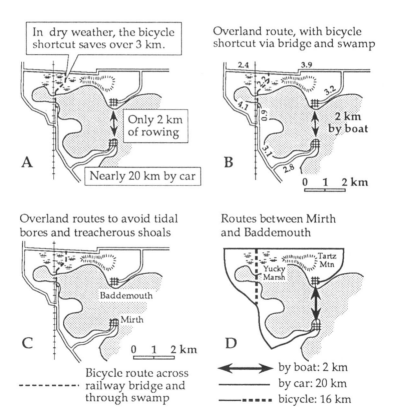

FIGURE 4.3. Four typical concept sketches illustrating a variety of alternate designs for a map comparing three routes.

text by assuring that the map contains all important facts and reveals relevant relationships, yet omits extraneous, potentially distracting details. These examples also illustrate how the use of type on maps affects the meaning of cartographic symbols and the interpretation of geographic relationships.

TYPOGRAPHY AS CARTOGRAPHIC SYMBOL

Although Bertin's theory of visual variables does not address type directly, labels on maps serve as primary or secondary cartographic symbols that reflect the six retinal variables discussed in the previous chapter.[7] Retinal variation allows labels to code features by kind and class as well as by name. For example, when the label "Santa Barbara" attaches a unique name to a point symbol that represents a medium-size city, the size and style of the type reinforces the primary graphic code of the point symbol, if the map key indicates that labels of this size and style represent cities with populations of 100,000–500,000. Knowledge of the retinal variations of type is important, because cartographic labels should reinforce, not contradict, the map's symbology. Type also reflects Bertin's two locational variables, because viewers tend to associate labels with nearby symbols. In addition to relating type to the visual variables, this section introduces basic terms useful in specifying type and examines the use of type as a primary cartographic symbol and as a tool for labeling places, streets, and other features named on the map.

Map authors need to know a few terms widely used in graphic arts and electronic publishing. A *typeface* is a type design

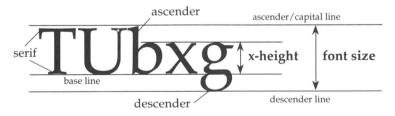

FIGURE 4.4. Font size is the vertical distance between horizontal lines slightly beyond the ascenders and descenders of lowercase characters, and x-height is the vertical distance between the base line and a line across the top of the *x* and other lowercase characters without ascenders.

identified by a name, such as "Palatino" or "Times," and a *font* is a full set of characters of a specific size for the typeface, including all capital, small capital, and lowercase letters, numerals, punctuation marks, and commonly used accents and symbols. Capital letters are also called *uppercase* characters, and letters not capitalized are called *lowercase*. The size of type is specified in *points*, one printer's unit equal to 0.0138 inch, or slightly less than 1/72 of an inch. Thus, 12-point type is approximately 1/6-inch tall. *Font size* is roughly the vertical distance between the lowermost reach of descenders on lowercase letters such as *g* and *j* and the uppermost reach of capital letters and the ascenders of lowercase letters such as *b* and *h*. As Figure 4.4 illustrates, font size generally includes a small amount of clearance for accents, decorative flourishes, and certain special symbols. The vertical distance between the base line and a horizontal line across the top of lowercase letters without ascenders, such as *a*, *o*, and *x*, is called the *x-height*. For fonts of the same size, typefaces can vary significantly in x-height, and when lowercase letters are used, x-height is a better indicator of legibility than font size. Most typefaces have *serifs*, short thin projections at the ends of main strokes on letters and numerals. Sabon, in which this book is set, has serifs, but comparatively plain *sans serif* typefaces, such as Helvetica, are sometimes used for cartographic labels. Within a typeface, the principal variations in style are *boldface* and *italic* type, which stand out from the normal style, called *roman*.

Type and the Retinal Variables

The alphabets, sizes, and styles available in a typeface provide the map author with a variety of retinal variables to portray qualitative or quantitative differences among features identified by name. As with the point, line, and area symbols examined in chapter 2, type favors some retinal variables more than others. Figure 4.5 illustrates that there are straightforward matches for orientation and size, a somewhat forced match for value, and a rich assortment of variations in shape. This attempt to match common typographic variations to Bertin's six retinal variables omits hue, because legibility requires a strong contrast between a label and its background.[8] For the light backgrounds of printed maps and light-gray video screens, black type is markedly more effective than even red type, although white and yellow are almost equally effective on video maps with a black background. Figure 4.5 also omits texture, because variation in the spacing of letters interferes with reading and is a poor way to portray either quantitative or qualitative differences. Moreover, although the value variations in Figure 4.5 suggest a number of strategies for using darker type to emphasize important features, these examples also vary in shape and size (thickness) and do not form a progressive series. Imagesetting equipment can generate type on a true value scale ranging progressively from white to black, but labels varying in graytone are likely to be illegible or difficult to distinguish from each other.[9]

A few straightforward principles guide the selection of type for effective cartographic labels. The first rule reflects Bertin's theory of retinal variables: Use size to represent differences in quantity or importance and shape to represent differences in quality. Inside the mapped area, labels varying in size might differentiate among categories of cities, based on population size, and among nations, states or provinces, counties, and minor civil divisions. Outside the mapped area, labels varying in size can reflect the relative importance of titles, subtitles, labels in the key,

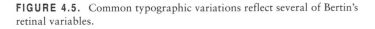

Size	*Value*	*Shape / Typeface*
7-point type	Medium	Palatino has serifs.
9-point type	*Medium italic*	Times, with a slightly smaller x-height, is more condensed.
10-point type		
	Boldface	
12-point	<u>Underlined</u>	Helvetica, with an even larger x-height, is sans serif and less condensed.
18-point	Reverse	

Orientation	*Shape / Style*
Horizontal	Roman *Italic* **Bold** ***Bold italic***
Curved Angled Vertical	ALL CAPITALS lowercase
	SMALL CAPITALS Initial Capitals

FIGURE 4.5. Common typographic variations reflect several of Bertin's retinal variables.

statements of scale, and notes about sources. In contrast, variations in style can differentiate labels referring to political units, cities, hydrographic features, and important groups of features, whereas a common size and style of type can reinforce the map's symbols by suggesting similarity among features in the same category. By convention, cartographers commonly reserve italic type to represent streams, rivers, lakes, oceans, and other hydrographic features. Lowercase type with initial capitals is convenient and effective to use for the names of cities and contrasts with uppercase type used for the names of states or provinces. Careful examination of commonly used topographic and regional maps might reveal other useful conventions for maps describing a landscape or a region.

Labels should always be easy to read. Legibility requires that uppercase type not be smaller than 6 points and that lowercase type not be smaller than 7 or 8 points, depending on the x-height and compression of the typeface.[10] For page-size maps in a book or scholarly journal, place-name labels in 9- or 10-point type will usually be legible to older readers with poor vision. Authors must be aware that an editor or publisher could make some labels illegible by shrinking the map to fit a small page or a narrow column. When in doubt about the minimum size of illustrations, check with the publisher or a number of potential publishers.

Important differences in the size of labels should be readily apparent to the viewer. For example, the difference between 9- and 10-point type might be too subtle to differentiate small and medium-size cities. For type sizes of about 10 points, a difference of at least two points is preferable, if size differences are high in the map author's hierarchy of concepts and if type provides the primary visual scaling.[11] Although 9-, 10-, and 11-point type might usefully distinguish among small, medium, and large cities if place-name labels merely reinforce point symbols varying markedly in size or pattern, it would be appropriate to use 8-, 10-, and 12-point type for point symbols that offer little contrast.

Avoid needlessly large type, an excessive variety of typefaces, and other visual distractions. Neither titles nor place-name labels should scream at the reader like headlines in a supermarket tabloid. In addition to detracting from the map's overall appearance, overly large type distracts the viewer and interferes with the perception of patterns and trends. Excessive use of boldface or reverse labels and use of a variety of typefaces on the same map are also aesthetically awkward. A single typeface can provide considerable tastefully coordinated differentiation with its various sizes and alphabets (capital and lowercase roman, capital and lowercase italic, small capitals, and, in some typefaces, capital and lowercase boldface). Although street maps for cities and other large, highly detailed reference maps usually require more typographic variety, a single typeface is adequate for most expository maps. Moreover, maps with labels in the same typeface used for the text look as if they were designed especially for the book or journal. There is one notable exception—sans serif type might provide greater legibility and more reliable reproduction when numerous very small uppercase street-name labels must be squeezed into tight spaces.[12] Another useful guideline calls for consistency among maps in a series. A common labeling style allows the reader to more easily and accurately interpret later maps using subtle typographic codes learned from earlier ones.

Labeling Places and Features

Labels identifying places and specific features have two roles, naming and signifying categories. On a map with identical lines representing both railways and highways, labels such as "B. & O. R.R." and "Pratt St." would clearly distinguish among principal types of transportation features. In contrast, on a map with different line symbols representing railways and streets, the signifying role of these feature names is to reinforce the graphic code of the line symbols and, for railways, to associate a route or right of way with a specific company. Street labels such as "Poe Ct." and "Franklin Al." also provide the important linguistic connotation of width and activity, which differentiate courts and alleys from boulevards and parkways.

A label's association with a specific feature depends on both typographic coding and proximity. The exclusive use of italic type for hydrographic features and uppercase type for state or national capitals makes it easier for the viewer to relate "Scots Brook" to a meandering line or "Sacramento" to an encircled star. But association between a feature and its label depends largely upon proximity. Unambiguous proximity is especially important for point symbols and small area symbols that cannot surround their labels. As Figure 4.6 demonstrates, labels should be close, but not too close, to their corresponding point symbols. Figure 4.6 also shows several more-preferred and less-preferred label locations around point features.[13] In practice, graphic conflicts with other features or labels might make one of the less preferable locations a quite suitable compromise.[14] In cases of extreme crowding, a leader line can sometimes link a feature to a label with no suitable nearby location. In many instances of crowding and graphic interference, though, the map author must increase the map scale, use a detail inset map, or omit some features.

Centering and orientation can also link labels with features. For large area symbols, centering the label within the symbol can be useful, as can enlarging or expanding the type so that the label extends across a large part of the symbol, as for the historian's U.S.S.R. map in Figure 4.7. Orientation can be important too; a nearly vertical label fits better within the narrow vertical boundary of Chile than a horizontal label does, and a gently curving label conveniently represents Czechoslovakia's shape and meandering east–west trend.

Orientation is also important for linear features. Figure 4.8 illustrates how street names and river names can follow the

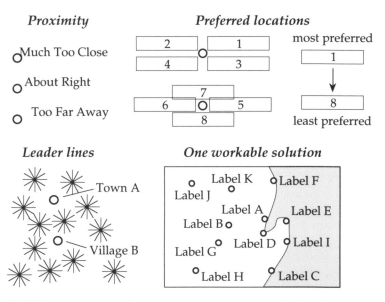

FIGURE 4.6. Strategies for using proximity to associate labels with point symbols.

alignment or trend of the feature. If possible, labels should read from left to right, with vertical labels reading upward toward the top of the page.[15] Curved labels are ideal for gently curving linear features; a convenient compromise for labels with two or more words is to keep a straight-line base for each part, but orient each component to the local direction of the feature. Yet as one of the examples illustrates, map authors should not accommodate a curved feature by hyphenating words and separating the parts.

Human-factors research indicates that place-name labels set in lowercase type with initial capitals are more easily recognized and more rapidly located than uppercase place-name labels.[16] There is a good explanation for this finding: lowercase letters provide more visual clues, especially for words with irregular upper and lower profiles caused by outwardly projecting capitals, ascenders, and descenders. Anyone who has ever had to read line after monotonous line of uppercase type will appreciate that "Albany" is far more easily identified than "ALBANY." Figure 4.9, which compares uppercase and lowercase labels for major cities in Alabama, demonstrates that 7-point lowercase labels with initial capitals are not only more legible but also tend

FIGURE 4.7. Strategies for using centering, orientation, and curvature to associate labels with area symbols.

to consume less space and cause less crowding than 6-point uppercase labels in the same typeface.

    Graphic marks between the letters of a label reduce legibility.
Strategies for avoiding graphic conflict between symbols and
type include repositioning the label, increasing the contrast between label and symbol, and blocking out interfering symbols
with an opaque background for the label. Obscuring important
parts of boundaries and other features can usually be avoided by

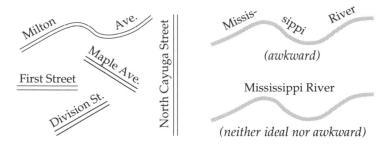

FIGURE 4.8. Strategies for using orientation to associate labels with line
symbols.

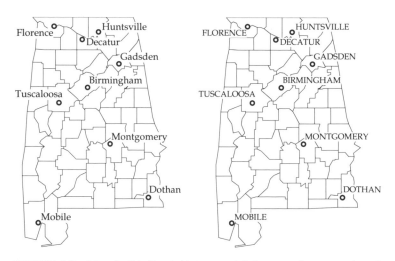

FIGURE 4.9. More legible 7-point lowercase labels occupy less space than 6-
point uppercase labels.

careful placement of labels, but increasing the contrast or using an opaque base is often necessary. Example B in Figure 4.3 demonstrates how large or medium-size bold type can provide sufficient contrast when a transparent label lies in front of a light-gray fine-dot screen. Smaller labels in plain type, as in example C in Figure 4.3, require an opaque background. Block-outs for labels crossing an area symbol should be centered symmetrically around the label and should be no larger than needed to keep the symbol away from the type. Where an opaque background does not interfere with line symbols, viewers usually can interpolate roads, boundaries, and other linear features across the label.

Although the spelling of all labels on the map should be checked carefully, place names warrant particular care. If a name has more than one spelling, the spelling on the map label must match the spelling in the text. Transliterating place names from other alphabets can be particularly troublesome, especially when a foreign government decrees a change (for example, China's replacement of "Peking" with "Beijing").[17] Political revolution and conquest can lead to name changes, often to honor a national leader (such as the renaming of St. Petersburg to Leningrad and back to St. Petersburg, and of Saigon to Ho Chi Minh City). Maps addressing an earlier period might benefit from use of the older name, especially if the text quotes records or literature of the time. But the current name should be noted somewhere, either in the text or on the map, in brackets. Map authors also should pay close attention to foreign words with accents and should watch out for inadvertent substitutions of zero for the capital oh or the numeral one for the lowercase el.

Labels as Symbols

Cartographic labels may be needed for unique names, such as "Philadelphia," that usually apply to a single place on the map, for names with both unique and generic parts, such as "Schuylkill River," and for generic names and abbreviations identifying two or more instances of a particular kind of feature. For example, a map author using a small black rectangle to represent the location of each post office in a city might place next to each carefully positioned little box a label reading "Post Office," or simply "PO." Such descriptive labels use natural language to provide a readily decoded link between the cartographic point symbol and the feature type. Thus, the real symbol is not just the black rectangle but the rectangle plus its label. Because it is spelled out, quite literally, the symbol need not appear in the map key.

Linguistic conventions support unambiguous abbreviations such as "PO." But when a comparatively vague alphabetic code such as "P" does not appear in the key, the reader must guess or infer from context whether the symbol depicts a post office, a parking lot, a police station, or a small park. Yet even an ambiguous single-character abbreviation can be a useful mnemonic, readily learned with a single inspection of the map key.

Physical scientists confronted with a large number of climatic types, geologic formations, or soil categories often find that abbreviations are the only solution. Some detailed soils maps are presented as a series of photographs annotated with boundaries and abbreviations and are bound at the back of a large book that itself serves as a detailed key; they would be impracticable without such abbreviations as "BnC" to represent "Bath–Nassau Complex, 8 to 25 percent slopes." Social scientists and humanists also find abbreviations useful as cartographic symbols, especially if the map must identify each area or point feature by its economic structure, land use, language, principal product, or dominant religion. An alphabetic code might include "D" and "Fn" to differentiate cities with a diversified economic structure from those specializing in financial services, or "AME" and

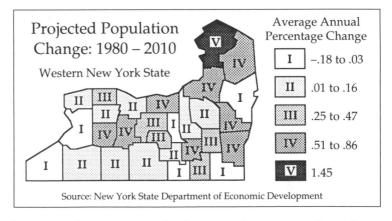

FIGURE 4.10. Roman-numeral codes representing map categories reinforce graytone symbols on a demographer's choropleth map of regional variation in the projected rate of population growth.

"Bp" to identify African Methodist Episcopal and Baptist congregations on a detailed map of African-American churches. Supplementing the map key with such parenthetical notes in the text as ". . . diversified cities (D) in the Northeast . . ." not only saves the reader a scan through the key but also calls attention to the map. Although a full key is necessary, either on the map or in a table, parenthetical references to important feature categories provide an effective link between the map and its associated text.

Alphanumeric characters can represent quantitative as well as qualitative differences. An anthropologist or rural sociologist using the letters A, B, C, and D to signify four progressively worse categories of farmland invokes the letter-grade metaphor instantly familiar to North American readers. The author can strengthen this link, as well as accommodate readers educated elsewhere, by using these letters directly, in discussing, for example, "category C farmland" or "prime agricultural land (A)." Roman numerals provide another convenient quantitative alphanumeric code, which can be reinforced by parenthetical references in the text, such as ". . . the low-growth counties (I, II) . . . ." As Figure 4.10 illustrates, use of roman numerals may bolster perception of a progressive series of graytones, as well as minimize the need for glances at the map key. Although overprinted alphabetic codes can diminish the perception of pattern by interfering with graytone area symbols, on maps printed in color they might serve viewers with impaired color vision.

Perhaps the most complex typographic symbology employed in the shumanities and social sciences is to be found in maps of linguistic patterns. Figure 4.11, which shows variation across Connecticut in the pronunciation of "New York," is part of a much larger map in the three-volume *Linguistic Atlas of New England*.[18] The atlas, based on field work carried out in the early 1930s, depicts regional variations in folk vocabulary and pronunciation. An elaborate key at the front of each volume describes the complex phonetic alphabet used to represent subtle yet important differences among local accents. This particular example shows the west-to-east transition in pronunciation of "New York," from a New York City–like accent to a more New England–like intonation that drops the *r* in "York." Readily decoded by linguists, these arcane symbols reflect the cumulative influence of important regional and local patterns of information flow and migration, which have led to the development of subtle but revealing dialect areas. This example also demonstrates how type and other graphically encoded natural language can be organized into an accessible cartographic format to promote scholarly communication.

LANGUAGE AND CARTOGRAPHIC COMMUNICATION

The words in the title and other labels link the map to the author's text and thereby help integrate with the rest of the book or article the patterns and relationships portrayed by the map's point, line, and area symbols. Titles and labels play the important role in cartographic communication of using natural language to impart meaning to graphic symbols. Effective cartographic communication requires that care and precision be used in the wording of titles, subtitles, source notes, caveats, and labels describing symbols listed in the key.

FIGURE 4.11. Map using a phonetic alphabet to show differences across Connecticut in the pronunciation of "New York." Maps in the *Linguistic Atlas of New England* also show rivers and ridges that might limit interaction between neighboring dialect areas.

Map authors need to be aware of the visual variations of type, examined earlier in this chapter. In addition to coding qualitative and quantitative differences among geographic features, appropriate typographic variation can promote legibility, expedite the search for features and places, and contribute to the map's aesthetic appearance. Labels identifying places and features also serve as cartographic symbols that supplement or reinforce the visual variables of the map's point, line, and area symbols. Labels ought never contradict a symbolic code they are intended to reinforce, nor should they distract the reader from the interpretation of geographic patterns and relationships.

Because titles and other labels are integral elements in map layout, their geometric arrangement affects their selection and should be integrated with the act of writing. Even when much of the design and all of the execution of the map occurs later, the author should at least sketch a rough version of the map, at the approximate publication size, while writing the related section of the text. Producing a rough-draft map is useful in forcing the author to think about the map's hierarchy of concepts and in judging and responding to the limitations of format. It may lead to identifying a need for an additional map to accommodate either more information than a single map can easily handle or a different treatment in another part of the manuscript.

A truly effective expository map supplements and complements an author's words. Expository cartography works best when it is integrated with expository writing, under the early creative control of the scholarly writer. There is a good reason, then, for the writer to be the map author, or at least for the scholarly author to have minimal knowledge of the principles and practice of cartographic communication.

CHAPTER

5

Cartographic
Sources and
Map Compilation

EFFECTIVE EXPOSITORY CARTOGRAPHY REQUIRES reliable sources. Like most intellectual discourse, after all, scholarly maps are at least partly derivative. Historical maps showing boundaries and roads almost always rely on one or more cartographic sources, and even maps describing patterns extracted through statistical analysis of geographic data need accurate outline maps showing boundaries of cities, census tracts, states, or other locational units. The field-oriented social scientist who maps primary observations requires a reliably detailed base map that positions with precision symbols representing routes, site boundaries, orientations, locations, and other spatial phenomena. The humanist concerned with a place or region must locate, evaluate, and interpret cartographic sources, resolve discrepancies, integrate maps with other forms of information, and possibly even reproduce maps as descriptive illustrations or significant artifacts. Because expository cartography is much more than designing and drawing maps, the researcher who recognizes early the need for maps can be efficient and effective in identifying and using appropriate sources.

This chapter offers a broad sense of what to consider, where to look, and what to look out for. Cartographic works relevant to the humanities and social sciences are so massive in number and so varied in theme and geographic focus that an entire book, much less a single chapter, could not adequately describe the range and variety of potentially useful sources of cartobibliographic information. Most readers are already—or soon will be—more aware than I of important bibliographies in their own fields. Since the topic is too important to ignore, my compromise strategy is to promote awareness of the problem and to outline in general terms how one searches for cartographic sources. The treatment includes a few specific examples, which may prove directly useful to some readers. The second topic in this chapter is the often-thorny issue of copyrights and permissions; again the problem is outlined and some general advice is offered. The chapter concludes with an examination of the mechanics of map compilation, the need to evaluate sources, and the question of when, whether, and how to use facsimile reproductions.

SEARCHING FOR CARTOGRAPHIC INFORMATION

A major part of knowing where to look is knowing what to look for. For example, a researcher exploring some aspect of mid- and late-nineteenth-century America could probably benefit greatly from consulting relevant county atlases. Between 1860 and 1910, a small number of commercial firms published large, detailed, and generally accurate atlases for most of the counties of the Northeast and Midwest, as well as for most of the nation's major cities.[1] Figure 5.1 illustrates the level of detail typically found, on a map of Eagle Bridge in an 1876 atlas of Rensselear County, New York. Because folk artist Grandma Moses [1860–1961] lived and painted there, this map might hold useful insights for an art historian interested in primitive American art. Similarly, an urban historian or sociologist concerned with the evolution of a particular city should be aware of the thousands of fire-insurance atlases produced for major American cities between 1870 and 1930, as well as aerial photographic surveying,

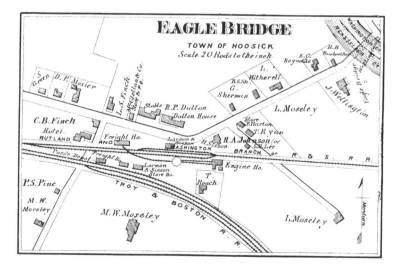

FIGURE 5.1. Example of a village map in a nineteenth-century county atlas. Because the map was reduced slightly, the printed scale is no longer accurate.

beginning in the early 1920s.[2] One learns about such sources as obsolete county atlases and insurance maps by reading the work of other scholars who have used them or by delving into the history of cartography for the period and country of interest. This section begins by listing and describing some common cartographic source materials of use to scholars and then moves to a concise discussion of cartobibliographies, map collections, and sources of information on the history of cartography.

Principal Types of Cartographic Sources

Awareness of the variety of contemporary cartographic products is helpful, whether the scholar is planning a research trip abroad or a half-day visit to the map department of a nearby research library. Knowing what might be available for the place, period, and phenomenon of interest can suggest what to look for and can lead to informative, serendipitous discoveries. If you have never explored the range of materials in a research library's map collection, do so soon. A good map department has all of the contemporary materials described in this section and many of the cartobibliographic aids discussed in the next section. In examining the more readily available types of maps, atlases, and other cartographic products, this chapter also notes important distinctions between government and commercial maps and between thematic and general-purpose reference maps.

First-time visitors to map collections often marvel at the massive flat-map storage cabinets. Many, if not most, of these cabinets hold "depository items" given to the institution by the federal government and available to the general public for inspection.[3] Depository maps are mostly large-scale, detailed topographic maps published by the National Mapping Division of U.S. Geological Survey. The depository collection usually also includes detailed coastal charts from the National Ocean Survey, various large- or intermediate-scale maps of geology, land cover, and water resources, and small-scale maps by the Bureau of the Census and other federal agencies. The Geological Survey also publishes small-scale maps of individual states, showing political boundaries, cities and towns, major transport routes, and principal physical features.[4] The Central Intelligence Agency publishes small-scale reference maps of foreign countries.[5] The Government Printing Office sells CIA maps and atlases, whereas the Geological Survey, through its National Cartographic Information Center, sells its own maps and provides information about maps by other government agencies.[6] In addition to helping you find specific maps, the map librarian can show you publishers' catalogs and provide addresses and other information about many public and private map producers, domestic and foreign.

Catalogs only hint at the wealth of data inside the map cabinets. Open a few drawers and see how topographic maps use symbols to represent the land surface, political boundaries, transport routes, streams, and shorelines, and also identify by name populated places, important features, and other landmarks. Note in particular how the Geological Survey uses color coding to differentiate layers of similar geographic features combined on a single sheet of paper: blue for water features; green for woodland; brown for contour lines and other terrain symbols; red for some roads and labels and reddish tint for built-up areas; black for most labels and for what cartographers call "culture," that is, roads, railways, buildings, and other constructed features. On some sheets, a sixth color, purple, points out new features or changes not shown on the previous edition. Because standardized Geological Survey topographic maps must support two distinct needs—economic development and national defense—they show many basic features useful for compiling new maps.[7]

Four series of information-rich topographic maps provide various levels of detail for nonoverlapping quadrangles bounded by meridians and parallels. A typical "7.5-minute series" map at 1:24,000 or 1:25,000 fits onto a 27- by 22-inch [68.6- by 55.9-cm] sheet and covers an area 7.5 minutes of latitude from south to north by 7.5 minutes of longitude from east to west.[8] An older, discontinued series published at 1:62,500 or 1:63,360 uses a quadrangle 15 minutes on a side, whereas a newer series at 1:100,000 employs a quadrangle covering 30 minutes of latitude by 1 degree of longitude. The 1:250,000 series has a quadrangle spanning 1 degree of latitude and 2 degrees of longitude. Figure 2.1, used earlier in this book to describe how map scale affects the level of detail, compares topographic maps showing Walden Pond at 1:25,000, 1:100,000, and 1:250,000. The Geological Survey also publishes a series of 1:2,000,000-scale National Atlas Regional Reference maps covering the entire country, including Alaska, with twenty-one overlapping sections.

Index maps show the area covered by each quadrangle in a map series. Every quadrangle has a unique name reflecting an important place on the map and the states within the quadrangle. The "Concord, Mass." 7.5-minute quadrangle, which includes Walden Pond, lists but one state in its name, whereas the geographically much broader "Boston, Mass.; N.H.; Conn.; R.I.; Maine" 1:250,000-scale quadrangle lists five. The Geological Survey publishes separate index maps for each state in its 7.5-minute series and separate nationwide index sheets for the 1:100,000 and 1:250,000 series. It also includes index maps in a series of catalogs listing the most recent edition of every Geological Survey map covering all or part of each state. Figure 5.2, a portion of a hypothetical typical index map, shows a grid of meridians and parallels, with each quadrangle identified by name; it also shows a highly generalized base map on which important places, political boundaries, rivers, and transportation routes provide a geographic frame of reference. Britain's Ordnance Survey, the Surveys and Mapping Branch of Canada's Department of Energy, Mines and Resources, and other foreign mapping agencies publish similar index maps.[9]

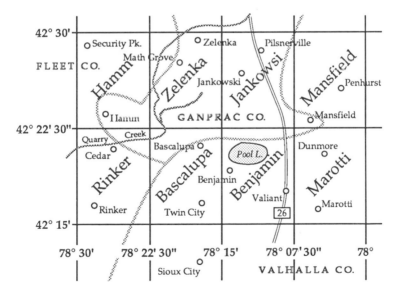

FIGURE 5.2. A index map showing the names of 7.5-minute quadrangles. Index maps often present quadrangle names and boundaries in solid black, to stand out from place names and geographic features printed in a visually recessive tan, blue, green, or gray.

Because topographic maps become less accurate as landscapes change, survey agencies produce new or revised editions, usually about every ten to twenty years. Yet old maps are still useful, especially for scholars concerned with past landscapes or with change itself. For example, Figure 5.3, an excerpt from a 1898 topographic map, describes for the local historian or industrial archaeologist the vast areas around Syracuse, New York, once covered by salt sheds, which had movable roofs to protect the massive solar-evaporation pans from rain.

As the number of new maps and revised editions has grown and as older maps have disappeared or become fragile, map libraries have turned to microfilm, microfiche, color slides, and other microformats to provide useful images and fill gaps in the collection. Microform publication of all Geological Survey topographic maps assures access to a more complete set of maps than the researcher is ever likely to find in the flat map cabinets. Microfor-

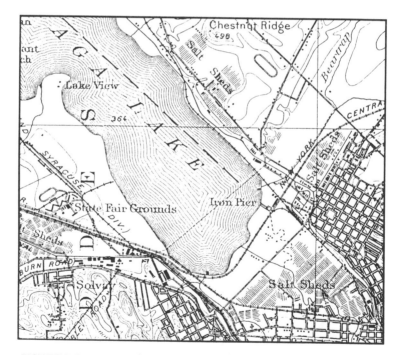

FIGURE 5.3. Portion of an early topographic quadrangle map of Syracuse, New York.

mat also allows map libraries to acquire inexpensive yet comprehensive sets of fire-insurance maps, nineteenth-century county atlases and land ownership maps, aerial photographs and satellite images, and foreign maps. Extension of the map collection through microform is particularly helpful because the United States did not undertake a serious, systematic program of detailed mapping for interior, noncoastal lands until 1882, shortly after Congress established the Geological Survey in 1879. Although a few states had mapping programs before then, the scholar usually must rely on the work of private, commercial cartographers and map publishers for maps of earlier periods. Black-and-white microfilm forces the map reader to rely more heavily on labels, context, and the shape of symbols. But micropublishing has the advantage of allowing scholars at many locations to use important maps otherwise available in few collections.

Although the government survey map is probably the best cartographic base for compiling a detailed, large-scale map, American topographic maps do not name most streets and are often out of date. Thus, a useful supplement to the topographic map is the commercial indexed street map. Many library map collections include inexpensive, contemporary indexed street maps for major world cities, as well as for smaller cities in the region, and store them folded in vertical files, rather than flat in a map case. Libraries periodically replace torn or obsolete street maps; so street maps more than a couple of decades old are more likely to be found in a local historical society or in state archives, if a curator has recognized their value and made a deliberate effort to unfold and preserve them.

Commercial map publishers, scholarly institutes, and private individuals annually produce tens of thousands of separate map sheets, and map librarians have developed extensive networks to help identify and acquire works relevant to their collections and clientele. It is not easy to catalog and store these diverse map sheets. Many librarians prefer vertical files for smaller, page-size maps, which are easily lost among the larger specimens in flat-map cases designed to hold sheets four feet or more across. Because the author-title-subject method of cataloging books does not accommodate maps well, map librarians have developed a variety of strategies for describing maps according to principal focus, scale, area covered, features included, author or compiler, date of the information, and date of publication.[10] Although most map libraries organize their commercial and nonseries government maps by subject for world maps and by region for other maps, systems vary, and the efficient researcher will either ask the librarian or read whatever guides are provided. Even if a library allows patrons free access to its map cases, browsing through map drawers is never as easy or productive as scanning book shelves.

Atlases are another significant part of a library map collection. They can be found in the reference room, the stacks (among the geography books), the map collection, the rare books department, and even the government documents section. Some are stored vertically and interfiled with the library's general book collection, whereas others are stored flat on deep shelves designed for large books. Older, historically valuable works, which often have brittle paper and crumbling bindings, are kept in the library's rare books or special collections department.

Although most of us think of an atlas as a book of maps, the term more appropriately describes any coherent collection of maps that more or less systematically addresses a theme, a region, or perhaps the entire world. While most atlases are bound and cataloged as books, the massive and detailed *Atlas of Switzerland* and a few other important atlases have been published in post-hole binders or large flat boxes, so that sheets can be issued over many years and replaced as needed.[11] The value of the atlas lies not in its similarity to a book, but in its comprehensive treatment of a subject or region and in its consistent use of symbols and projections to help the user relate one map to another.

World atlases are a standard part of a good reference collection. World atlases from the National Geographic Society, Rand McNally, the *Times* of London, and other prominent atlas publishers are commonly included. In contrast to a world reference atlas, some atlases focus on a single nation, state, or region. Others address a specific topic, occasionally for the entire world but usually for a country or region. These atlases are rich in *thematic maps*, which focus on a specific distribution or theme, rather than on general-reference features such as boundaries, place names, and major transport routes. Figure 5.4 is an example of a thematic map; it could offer regional economists and political scientists an informative snapshot of the relative abundance of small farms in the southern Appalachians. Although commercial publishers produce most thematic atlases, government agencies often tap their own data and issue highly informative atlases addressing agriculture, disease, population, and other important themes.

Government publications such as the *Graphic Summary of the Census of Agriculture*, from which Figure 5.4 was taken, often wind up outside the map collection in the library's government documents section. Indeed, scholars can find many useful maps outside the map room. After all, maps are often included in books, periodicals, and newspapers. For example, *Geographic Notes*, a biannual publication of the U.S. Department of State, contains numerous maps, such as Figure 5.5, which accompanied a 1991 article on the geopolitical implications of Yugoslavia's ethnic diversity. Depending on the whim of the library staff, *Geographic Notes* may be found with maps, government documents, or general periodicals. Another example is Figure 5.6, a woodblock engraving of Baltimore, Maryland, published in the *New York Tribune* early in the American Civil War. Because its caption notes that the map was "kindly furnished for THE TRIBUNE by an officer of the United States Engineers, [and] exhibits very clearly the points in Baltimore which might be held by military forces whether for purposes of attack or of occupation merely," this newspaper map is of value to media scholars as well as to students of American history. Of course, locating maps employed as illus-

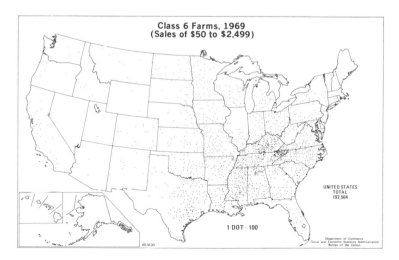

FIGURE 5.4. Example of a map in a thematic atlas.

trations is much more difficult than locating maps published in an atlas, as part of a map series, or as separate items. But as I discuss in the next section, bibliographic aids in the map library can point the researcher to some of these materials.

Map collections sometimes include printed outline maps that show shorelines and important boundaries, major cities, princi-

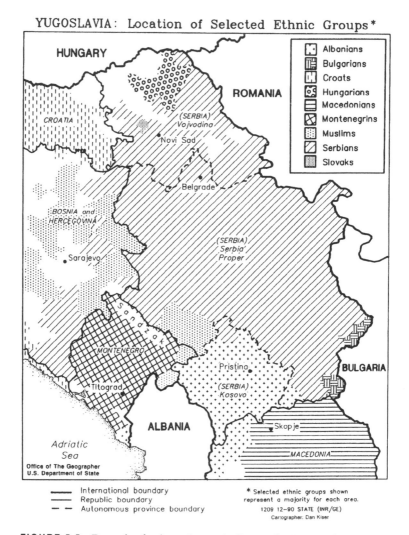

FIGURE 5.5. Example of a thematic map in *Geographic Notes*, a biannual publication of the U.S. Department of State.

pal rivers and lakes, and possibly some transport routes or a highly generalized representation of the terrain. Commonly sold in book and stationery stores, the outline map is designed as a base on which the researcher may plot information with marker pens or colored pencils. In some cases, adding type and a few carefully chosen press-on symbols yields a map suitable for publishing as an illustration for a book or journal article. Although outline maps may be photocopied or borrowed for use in map compilation, many researchers will find the library's holdings more useful for identifying publishers from whom fresh copies can be ordered.

Map libraries have adapted slowly to the growing availability of electronic cartographic products.[12] They are certain eventually to offer access to an electronic publishing system for displaying cartographic data and making publishable maps on a laser printer. In addition, they will provide accessible clip-art files for use with electronic drawing packages such as MacDraw. Figure

BALTIMORE, AND ITS POINTS OF ATTACK AND DEFENCE.

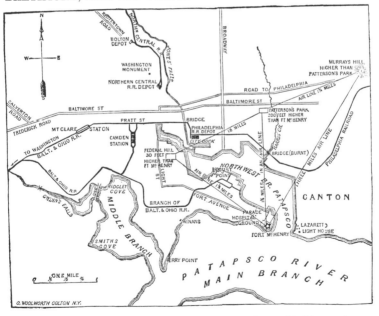

FIGURE 5.6. Map showing preparations for the defense of Baltimore that appeared in the *New York Tribune* on 30 June 1861.

5.7, produced by adding some information to a commercial clip-art map of Japan, preserving only selected places and labels, is the electronic equivalent of a map made by plotting symbols on a printed outline map. The electronically competent map library will also provide geographic boundary and city location files for use with map-making software such as WORLD (used to draw some of the projections in chapter 2) and ATLAS MapMaker (used for several statistical maps in chapter 6).

The few significant library collections of air photos have come into being because of a generous donation or an academic pro-gram active in land analysis, photogrammetry, or remote sens-ing. Most of the aerial and satellite imagery found in map collec-tions has been reproduced on paper by a printing press and consists largely of photographic quadrangle-format maps and color-composite satellite images distributed as depository items by the Geological Survey. But contemporary air photos provide useful details omitted on printed maps, and older aerial imagery can be well worth tracking down. As a case in point, Figure 5.8

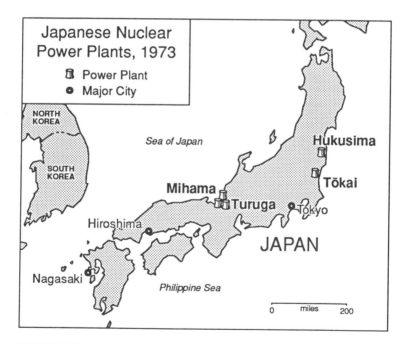

FIGURE 5.7. Map of nuclear power stations in Japan, 1973.

FIGURE 5.8. Portion of an early air photo of Syracuse, New York, covering area to the left and below the center of the region shown in Figure 5.3. The photograph reflects substantial change between 1898 and 1926. For orientation, focus on the portion of the old Erie Canal entering both illustrations from the left and on the multiple-track railroad just below the new Fair Grounds race track on the photo.

shows a portion of a 1926 air photo covering part of the area shown in Figure 5.3. I found this photo, not in the university map library, but in a collection of contact prints acquired in the mysterious past by my department's cartographic laboratory. It shows the then-new race track at the New York State Fair Grounds and the remnants of the road network serving the massive salt sheds. This photo and other images in the set would be useful to scholars interested in local history or industrial archaeology, and I hope to make them more widely available through the university library's special collections department.

Cartobibliographies and Major Collections

A general sense of what might be available can be a strong incentive to track down important maps or promising collections. To illustrate how the researcher might proceed, this section examines briefly a few significant or representative sources of several types: directories of map libraries, facsimile catalogs for particular map collections, indexes to maps on specific topics, indexes to maps in books and periodicals, and bibliographies on the history of cartography.

A directory of map libraries can help the scholar find a promising collection. Most directories are compiled from responses to mail questionnaires. They describe briefly each collection's size, specialities, and cataloging practices and provide addresses, phone numbers, and hours. Directories often note the collection's facilities and policies for tracing, photocopying, and photographing materials.

Directories tend to be worldwide, national, or local in scope. Because of language differences and the difficulty of international coordination, worldwide directories are rare. Professional organizations, especially the Geography and Map Libraries Section of the International Federation of Library Associations and Institutions, were particularly important in compilation of the *World Directory of Map Collections*.[13] To achieve global coverage, the *World Directory* was restricted to map collections with at least a thousand maps or atlases. The *World Directory* is particularly useful for scholars planning visits to the Third World and other foreign areas.

Because compilers addressing a single nation face a more straightforward task, national directories are more common and

more comprehensive than world directories. Researchers concerned with North American material can choose the *Guide to
U.S. Map Resources*, published by the American Library Association, or *Map Collections in the United States and Canada: A
Directory*, published by the Special Libraries Association.[14] Both
directories group entries by state and city, include an extensive
subject index, and present a variety of statistics describing individual collections. The *Guide* lists 975 individual collections
holding 285,000 atlases and nearly 37 million maps; *Map Collections* has 804 entries. Both directories are revised regularly.

National directories describe map collections in many foreign
countries. For instance, researchers can consult *A Directory of
U.K. Map Collections*, compiled with the help of the British Cartographic Society's Map Curators Group.[15]

Among the few cities with the number and variety of map collections to warrant their own directories are New York and
Washington, D.C. Ralph Ehrenberg's *Scholar's Guide to Washington, D.C., for Cartography and Remote Sensing* is an indis-
pensable planning aid for the researcher visiting Washington, as
well as a valuable source of addresses, telephone numbers, and
basic product information for anyone interested in contemporary maps, especially maps published by the federal government.[16] More comprehensive than most directories, the *Guide*
covers maps, charts, aerial photographs, satellite images, cartographic literature, and digital cartographic data available at the
numerous public and private libraries, archives, museums, federal agencies, professional societies, and other organizations in the
national capital. *A Guide to Historical Map Resources for
Greater New York* is smaller and focused on the era before 1950.
It lists forty-nine collections, mostly in the New York metropolitan area.[17]

A one-page listing in a directory cannot begin to describe the
Map Room of the British Library, the Geography and Map Division of the Library of Congress, the Map Division of the New
York Public Library, or any other large map library. Although the
only way to take full advantage of a map collection is to visit and
use it, researchers can gain a fuller sense of the depth and breadth
of these and other important collections by consulting the map
catalogs published for most of the important collections.[18] The
American Geographical Society collection is one example; its

catalog of maps, books, and atlases appeared in ten volumes in 1968, with supplements in 1971, 1976, and 1986. Although the map catalog of the British Library is set in type, most others consist of facsimiles of catalog cards assembled on a large frame in groups of two dozen or so and photographed as folio-size pages.[19] Because the catalog of a map library with a special focus or a nearly comprehensive collection is a valuable bibliographic tool, research libraries strong in cartography are likely to have at least a few facsimile catalogs from major map collections. Research libraries unable to obtain bound paper copies often acquire these catalogs in microform. Printed, microform, and electronic catalog supplements are available, so a catalog first produced in the 1950s or 1960s can be nearly as current as the one in the distant map collection's own reference area.

Researchers seeking maps as data for compilation can consult a variety other sources found in a good research-oriented map library. Several cartobibliographies are especially useful in tracking down contemporary, late-twentieth-century maps and atlases. *Information Sources in Cartography* offers an overview of map production, map librarianship, and the variety of maps currently in use, and suggests numerous bibliographic sources, publishers, and mapping agencies the researcher might consult.[20] *World Mapping Today* discusses topographic and resource mapping for continents and individual countries, describes each country's important map series, provides graphic indexes for some intermediate-scale series, and lists sources of cartographic information within each country.[21] Prepared with the support of the International Cartographic Association, the *Inventory of World Topographic Mapping* describes individual countries' programs of large-scale mapping and includes facsimile excerpts of various nations' topographic maps.[22] *Kister's Atlas Buying Guide, National and Regional Atlases*, and *International Maps and Atlases in Print* are all useful yet somewhat dated sources of information about atlases.[23] *Bibliographic Guide to Maps and Atlases*, an annual summary of new acquisitions at the Library of Congress and the New York Public Library, is an up-to-date guide to new maps and atlases.[24]

Research catalogs and general bibliographies can also point the researcher to specialized cartobibliographies covering a variety of topics, usually historical and mostly focused on a particular region or map collection. Examples of cartobibliographies on

North American history include *Picturing America, 1497–1899*; *Georgia at the Time of the Ratification of the Constitution*; *Maps and Charts of North America and the West Indies, 1750–1789*; and *Maps and Charts Published in America before 1800.*[25] Staff at the Library of Congress have prepared annotated bibliographies for a number of important American cartographic themes, including Civil War maps, fire-insurance maps, land-ownership maps, and panoramic maps of cities.[26] Other bibliographies are more international in scope, such as *Maps of the Holy Land: Cartobibliography of Printed Maps, 1475–1900.*[27]

Unfortunately, many useful maps lie uncataloged in books and periodicals. Although bibliographic guides to the general literature on a topic sometimes note whether particular references include maps, researchers interested in maps incorporated in larger works have few indexes to help them. The best source for maps in early- and mid-twentieth-century publications is the American Geographical Society's *Index to Maps in Books and Periodicals*, published in ten volumes in 1968, with supplements in 1971, 1976, and 1986.[28] Like the Society's *Research Catalog*, the *Index* is a facsimile reproduction of a card catalog. A potentially promising source for older material is the series of guides published by Massachusetts bibliographer David Jolly. Its *Maps of America in Periodicals before 1800*, for example, has 465 entries indexed by place, publisher, and periodical.[29] Jolly offers advice on the cartographic proclivity and relative rarity of carefully selected periodicals and mentions specific collections where these periodicals and their maps may be consulted.

Some guidance may be useful as to how to cite a map used as a source or reproduced as a facsimile. Cartobibliographies often provide useful examples. If the map is included in another publication, cite the book, atlas, or journal in which it appears and include the page and the figure, map, or plate number. If your publisher's style specifications do not include a standard citation form for sheet maps and other cartographic materials, adapt the recommended style for books and monographs. When no author is specified for the map, as for most government and commercial maps, treat the map's publisher as its author and omit a redundant reference to the publisher. For series maps, three of which are cited as sources for Figure 2.1, treat the sheet name as the title and include the scale or series name. Also include the publication date and any other information useful in distinguishing the version consulted from other editions of the map. If a map is rare or difficult to find, include in parentheses or brackets the name and location of the archives in which the map was found, as well as a catalog number, accession number, or other information to help another scholar who might want to examine the map. If the map is reproduced as a facsimile, mention the size of the original or the reduction factor, if any. When an original edition printed in color is reproduced in black-and-white, provide a short description of the original. If the description seems excessively long for a bibliographic citation, consider using an information note to the text.

Compiling from maps without understanding why and how they were produced and published can be risky. Are newer maps always more accurate, for instance? Have more developed areas always had better maps than less developed areas? Two works that can help scholars understand how technology, politics, or intellectual climate might limit the accuracy of cartographic sources are the *Bibliography of Cartography*, maintained by the Geography and Map Division of the Library of Congress, and the *History of Cartography*, edited at the University of Wisconsin by geographers J. B. Harley and David Woodward and published by the University of Chicago Press.[30] Another facsimile card catalog with periodic supplements, the *Bibliography* includes references both to historical studies of map making and contemporary articles on mapping from the late nineteenth century onward. It is a useful complement to the massive and unprecedentedly thorough *History*, the first volume of which appeared in 1987; two decades may be required for completion of its six volumes. These two works and *Imago Mundi*, a periodical published by the International Society for the History of Cartography, can help researchers identify and understand important pre-1900 maps and atlases.

COPYRIGHT AND PERMISSIONS

Copyright regulations confuse many scholars who want to extract information from existing maps. Although publishers' style books advise on how many words one may quote from a book without asking permission, they offer no equivalent guidelines for plucking features or facsimile excerpts from maps.[31] Equally mysterious are international differences in copyright protection;

for example, an author can use U.S. Geological Survey topographic maps as sources free of charge, yet cannot publish a map compiled from equivalent British sources without first obtaining permission and sometimes paying a small fee.[32] Researchers are often uncertain about exactly what a map copyright safeguards. After all, what creative, original material is there to protect when the copyrighted map from which we want to borrow was itself compiled from previous maps in the public domain? Because most American maps originate with information to which we as citizens are already entitled, the map challenges the legal notion of intellectual property.

In this section I offer a pragmatic route through the copyright maze by examining relevant legal principles and offering general advice on what the scholar may do, should do, and may not do. I also look at the copyright provision for critical comment and review of visual material, even over objections of the copyright owner, and provide a sample request for permission that several of my publishers have found acceptable.

The Enigma of Cartographic Copyright

Understanding how copyright regulations affect map authors requires an understanding of what can be copyrighted and what kinds of copying the law prohibits. To begin, the law requires that the copyrighted work be original, creative, and fixed as an artifact on paper or film, in computer storage, or in some other tangible medium.[33] No one can copyright an idea such as a method of symbolization, a fact such as the shape of a river, or a government survey in the public domain for free use by all citizens. But a map maker can copyright a specific map that portrays coastlines, rivers, and other linear features with unique, carefully crafted line symbols. In short, a map's originality and creativity are in its generalization and symbolic portrayal of geographic features and in its selection of details. The law protects the fixed graphic expression of these features, not the information on which their representation is based.

Infringement of a map copyright is somewhat like excessive quotation from a copyrighted book: although it is legally acceptable to describe what an author wrote and even to paraphrase portions, it is potentially actionable to use an author's exact words or to paraphrase excessively. Thus, to photographically reproduce all or part of a validly copyrighted map can be a clear case of infringement. And to reproduce a carefully traced, re-drawn version of a coastline can also be evidence of copyright infringement. Yet to redraw the coastline as a smoother, less jagged symbol can be a graphic paraphrase of the map maker's representation of the coastline and not a violation of copyright. Moreover, to trace from the same public domain source another map maker used does not infringe on the copyright, even though that representation and yours might be highly similar. Indeed, if both representations were exact replicas of the coastline symbol on a Geological Survey map, neither portrayal could have a valid copyright. After all, the intellectual property protected by the copyright act is not a set of geographic facts but the original and creative graphic representation of those facts.

Because copyright legislation rewards creativity, not effort, trying to compile a derivative map without infringing someone's copyright might seem like a game in which deliberate embellishment helps the winner avoid an opponent's trap. Yet thorough-ness, thoughtfulness, and other principles of good scholarship lead us around, if not away from, the legal pitfalls of cartograph-ic compilation. For instance, we violate no one's copyright when we compile from *primary* sources in the public domain, such as maps produced by the federal government and old maps for which the copyright has expired. We can reproduce these maps freely, as I have done in Figure 2.1, or redraw them as precisely as we choose. Or we can alter original artwork in the public domain by cutting it up, masking out some features, or adding others.[34] A map author can sometimes avoid copyright infringement by de-riving a new map from several different sources, some copyright-ed and some in the public domain. Street-map publishers often do this—they call it "editing the competition"—to avoid picking up "trap streets" and other spurious features inserted by wary competitors to establish grounds for legal action.[35] Scholars should be as cautious, to avoid unintended as well as any deliber-ate errors.

Because knowing when a map was created—when it became "fixed in tangible form"—is especially important, authors should be wary of hand-drawn or undated maps. Before Con-gress changed the copyright act in the late 1970s, common law protected manuscript maps, drafts, and other maps not yet pub-lished, and to reproduce one, it was necessary to obtain permis-sion.[36] Prior to 1978, publication removed this protection, and a map lacking a valid copyright notice in its first publication entered the public domain and could be copied freely. In contrast, the new law affords protection from the instant a work is created. Since February 1989, the law no longer requires the traditional copyright notice (such as "© HAMMOND INCORPORATED"), but publishers continue to use these notices to discourage infringement.

Copyright has never protected a work indefinitely. Because Congress has revised the law several times, the life of a copyright depends on several factors, including when the work was first published and when its author died. In general, the current law protects material copyrighted after January 1, 1978, for the life of the author plus 50 years.[37] In contrast, the law protects works copyrighted before 1978 for 28 years, 56 years, or as long as 75 years. The earlier law specified an initial term of 28 years, with the possibility of renewal for an additional 28 years.[38] However, during the 1970s, Congress extended the renewal period to as long as 47 years in some cases, for a total of 75 years. But this extension applies only for copyrights that were actually renewed: if the copyright of a work expired after 28 years, it passed into the public domain and can now be copied without restriction. Of course, ascertaining that a copyright was allowed to expire might require obtaining a letter from the owner or the owner's heirs, carefully examining the *Catalog of Copyright Entries* published by the Library of Congress, or ordering a copyright search by Copyright Office staff.[39] Map authors should also be wary of books and maps protected by foreign copyrights, which sometimes have longer terms than American copyrights.[40]

The 75-year rule is a boon for historians and others who might like to reprint a facsimile of an older map. When I wanted to include Elkanah Tisdale's famous 1812 Gerrymander map-cartoon in *Maps with the News*, I found a book on the history of cartoons and caricatures published in 1877.[41] Although I might have copied the Gerrymander drawing from any of several more recent books on early American history or political redistricting, I was able to find a sharp, detailed example without needing to worry about having to obtain permission and possibly pay a fee. *Mapping It Out* is being published in 1993, so I can safely reproduce here any map published in the United States before 1918 (1993 minus 75 years).

Fortunately for scholars interested in contemporary maps, the doctrine of fair use permits reproduction of material still under copyright for a range of purposes, including criticism, education, news reporting, research, and scholarship.[42] In limiting a copyright owner's exclusive rights, Congress recognized that some uses are not only in the public interest but so minimal as to make obtaining permission and paying a fee unwarranted. Section 107 of the Copyright Act of 1976 lists four criteria for determining whether copying constitutes infringement:

1. The purpose and character of the use, including whether such use is of a commercial nature or is for nonprofit educational purposes
2. The nature of the copyrighted work
3. The amount and substantiality of the portion used in relation to the copyrighted work as a whole
4. The effect of the use upon the potential market for, or value of, the copyrighted work.[43]

These statutory criteria are not specific, but both case law and legal scholarship on fair use suggest that the most important test of fairness is whether copying would have a harmful economic impact on the sale of a copyrighted publication.[44] It's not surprising, then, that suits by commercial street-map publishers against their competitors account for much of the litigation involving cartographic copyright.

Common sense must compensate for the lack of specific guidelines on fair use. Consider, for example, the black-and-white "reference map" sold at a downtown stationery store in packages of twelve. Although the map contains a copyright notice, tracing a few of its features for a map to be used in a journal article seems well within the scope of fair use, if you consult other sources and generalize appropriately.[45] It is not necessary to cite your sources if the information is not unique and is widely available. Making a dozen photocopies in order to experiment with various designs technically infringes the copyright because this act denies the publisher an additional sale, but since the purpose is research and the economic harm is slight, the dozen copies are probably acceptable under fair use. In contrast, adding your own symbols and type to the map and reproducing it in a book or journal might indeed infringe the copyright; the presumption is always that rights must be purchased unless fair use excuses the use. Yet if the package, the publisher's catalog, or non-photo-blue lines and place names on the map itself identify the work as a reproducible outline map, photographic copying or direct tracing is clearly legal, as long as you don't threaten the publisher's sales by manufacturing and selling outline maps. Indeed, copying is a fair, normal, and anticipated use of the product. Other examples of cartographic products that the seller assumes the buyer will use to create new maps are: the boundary files supplied with mapping software; electronic base map files, such as MicroMap's MAP Art (used in making Figure 5.7); and printed collections of cartographic clip art, such as the clearly titled *Asia Today: An Atlas of Reproducible Pages.*[46]

Fair use permits some unauthorized copying for critical, evaluative studies of maps in advertising and political propaganda and for critical reviews of maps and atlases.[47] Because copying can be essential for the effective review and criticism of persuasive, controversial, or otherwise noteworthy maps, fair use allows scholars to reproduce excerpts or a reduced image of the whole map even when the copyright owner denies permission. Of course, the scholar must conscientiously discuss the map as an artifact rather than exploit its geographic information; the provision for fair comment is hardly an excuse for saving the cost of producing your own map.[48] Moreover, fair use could never sanction the unauthorized reprinting of a copyrighted map to decorate a book cover or an advertisement. Book reviewers, literary critics, and art columnists regularly take advantage of the copyright law's support for review and comment, but, unfortunately, scholars studying maps rarely do so.

As a practical matter, you must consider how feisty the copyright owner might be and whether your publisher would stand behind you. Because publishers want to avoid costly litigation, they almost always require written permission to include a facsimile excerpt of a copyrighted work. But some publishers are less easily intimidated, and some copyright owners are less likely to litigate. The *New York Times* would be more willing than the Southwest Wyoming State University Press to risk harassment from the gun lobby or a large tobacco company, for instance. Yet even the SWSU Press might be willing to endure the comparatively impotent ire of the Toxic City Chamber of Commerce.

One final caveat: Although reproduction rights are different from access rights, comparatively rare maps found in archives could raise the issue of photographic copyright.[49] Although a map published in an 1850 book is well beyond copyright protection, the owner of the artifact, who photographs it for you, might place restrictions on your use of the photograph. Such restrictions could also apply to a machine copy, because few archives will let researchers make their own machine copies of rare, fragile maps. When you pay for the photograph or machine copy, you must make certain there are no strings attached. Generally, you are free to use the material unless specific restrictions are placed on reproduction of the copy as a condition of sale. In providing a copy without such restrictions, the institution has, in effect, granted access.[50] You then need only consider whether someone holds a valid copyright on the map. If the map is in the public domain, or if fair use applies, you may reproduce the map freely; otherwise, you need to obtain permission.

Obtaining Permission

Getting permission to reproduce a map usually requires only that you write the copyright owner and ask for it. Inside the genteel world of academic research, scholars normally request and receive permission when they want to reprint a map published in a book or scholarly journal. In most cases, this exchange of letters is merely a professional courtesy, and no fee is assessed. Any permission fees required are usually small and sometimes negotiable, especially for nonprofit use. When you are merely reexpressing cartographically unique facts or field observations reported by another author, fair use makes seeking permission unnecessary, even though good scholarship demands a proper citation. Among academics, a charge of graphic plagiarism is far worse than a questionable accusation of copyright infringement.

Nonacademic sources vary widely in their willingness to grant permission. Most charge nothing for a facsimile reproduction and ask only for a copy of the publication and, of course, a courtesy note in the figure caption or list of illustrations. Although a few copyright owners might seem greedy, most are flattered and cooperative. If the charge seems excessive, you must pay it, negotiate a reduction, substitute another map, or drop that particular example entirely.

Periodical and book publishers are usually helpful and reasonably prompt. A magazine that purchased only first-reproduction rights will send you the name and address of the artist. A publisher that didn't renew its copyright for a 1935 book will not only tell you so but will provide useful documentation for your permissions file. A good rule of thumb is to suspect that a map is still under copyright and to seek permission from the presumed copyright owner.

A form letter with a release form at the bottom of the page is the usual way to request permission. Send two copies, one for the copyright owner to keep and the other to be signed and returned in the stamped, self-addressed envelope you include. A few publishers use their own form and even specify the wording of the citation or credit line. But most will find your properly worded form letter acceptable.

Requesting permission to reproduce a map is little different from requesting permission to quote a poem or text from a book. Use a form letter similar to that in Figure 5.9, preferably on university letterhead or personal stationery.[51] My request identifies the material I want to copy, where it was published, and the rights I seek; describes the book or article I am writing and how the map will be used; and promises to give proper credit. I routinely ask for nonexclusive world rights, in all languages and for all editions. For clarity, I include a copy of the map or portion I want to reprint, and if the map appeared in a book, a copy of the book's title page and copyright page.

Your request for permission to reprint a map should describe how it might be altered. The copyright owner should be told, for instance, that you plan to reproduce the entire map as a reduced, page-size halftone or as a nonreduced, same-scale excerpt. If a color map is to be reproduced in black-and-white, say so and indicate what symbols or type might drop out. As a courtesy, indicate why you want to include the map and enclose a copy or sketch of the altered version and the caption you have drafted for it. These details are appropriate, because the law lets the copyright owner control how the work is reproduced.

A separate cover letter might be useful. Although telling the copyright owner you want to discuss the map as a particularly good example might encourage a prompt, positive response, never deliberately misrepresent your intentions. If you receive no response after a month, call or send a new cover letter with two copies of the original permission request and a copy of your original cover letter. Because the copyright owner might be difficult to locate and a refusal will require revising the manuscript, screen maps for possible use as facsimile illustrations while you do your research, set up your form letter early, and request permissions as you write your first draft.

```
                            [my letterhead]


[date]                                      Reference:


Rights and Permissions Editor
[publisher's name and address]

This is a request for permission to use [description of map]
included with [name or description of publication, edition,
and copyright date].  I enclose copies of the map and [the
publication's (if a book) title and copyright page].

I request permission to use this map as an illustration in a
book tentatively titled [my title], to be published by [my
publisher] in [19??].  My book is a study of [???], and I
want to include this map among several examples of [???].

If you are the copyright holder, may I have permission to
reprint in my book the material described above?  I am
requesting nonexclusive world rights to use this material
for all editions of my book, in all languages.  I will cite
the source and mention that the material is used with
permission.

If you are not the copyright holder, or if the copyright has
expired without renewal, please so indicate.

Thank you for your consideration of this request.  Please
sign and return one copy of this letter.  A duplicate copy
is enclosed for your files.

Yours sincerely,

[my name]
[my title]
- - - - - - - - - - - - - - - - - - - - - - - - - - - - - -
The above request is approved on the conditions specified
below and on the understanding that full credit will be
given to the source.

Approved by:                                Date:
```

FIGURE 5.9. Example of a letter requesting permission to reproduce a map.

THE MECHANICS OF COMPILATION

Because cartographic sources vary in scale, time, subject matter, projection, symbolization, and reliability, map compilation is not a single straightforward process. In simple cases, the map author may need to extract and generalize a few basic features from a public-domain base map or add to it some new features identified by address or coordinates. In other cases, a composite map may require a range of sources, some covering only part of the area, some providing but one or two features. Occasionally the map author must fit together two or more adjoining sheet maps or attempt to resolve differences in content as well as geometry among maps produced at different times by different publishers. This section outlines basic steps the map author should follow to compile comparatively complex maps. It then examines the special requirements of cartographic facsimiles.

Evaluating the Usefulness of Sources

When compilation requires several sources, the map author needs a systematic strategy for evaluating the available data. If the sources cover different parts of a wide area, a working source-coverage map is useful for showing overlap and pointing out zones where coverage is missing, not corroborated, or based only on a single questionable source.[52] If there is much geographic overlap, a table with rows for each source and columns listing the projection, scale, content, date, publisher, and estimated reliability can organize the data for systematic evaluation. Noting which maps cover comparable territory or subject matter provides a useful checklist for the cross-comparisons important in evaluating accuracy.

Evaluating the map scales listed in the table can point out unsuitable or inappropriate sources and can indicate how the scales of the key sources might limit the scale and detail of the compiled map. To avoid error, cartographers generally compile from larger to smaller scales, that is, from more detailed sources to a less or equally detailed compilation. Because map viewers expect or assume comparable and consistent levels of detail, the inclusion of features from a much smaller-scale source could be confusing or misleading. For example, a map at 1:2,000,000 is too vague a source for rivers, streams, and shorelines for a transportation map compiled at 1:50,000. The smaller-scale map lacks important streams, and its overly smooth features would not accord well with features compiled from more detailed sources. Yet, like most guidelines, the larger-to-smaller-scale rule requires flexibility and logic. When evaluating the suitability of sources, the map author must consider the compiled map's goals and scale of concepts. After all, if a 1:50,000 map is to focus on railways and major highways, a 1:100,000 map might show all the details about hydrographic features that the viewer really needs.

When looking for sources, the scholar also tries to identify a satisfactory compilation base. A reliable map covering the entire area and containing most of the needed features might provide the cartographic framework for efficient compilation. Transferring information onto a single compilation base map is easier and generally more likely to produce accurate results than attempting to work out a running compromise among several sources. Ideally the compilation base is either a map in the public domain or a map the researcher has already developed. (If the book or article will need several maps covering the same area at the same scale, a standardized map base not only simplifies compilation but gives the reader a familiar geographic framework.) The ideal compilation base is also similar in scale to the finished map and requires little further generalization: there is little point in compiling from smaller-scale sources onto a larger-scale compilation base and then having to generalize the compiled drawing for publication. If some features must be transferred "by eyeball," a grid of meridians and parallels can be useful. If no appropriate map covers the entire area, the next best solution is a series of quadrangle maps that fit together easily because of a similar scale and projection.

Although large-scale maps are more geometrically precise than smaller-scale maps, scale is only one of several factors affecting a source's suitability. As discussed in chapter 2, some projections might be more appropriate than others to the map's goals and focus, because of properties such as the preservation of area, the conservation of local shape, and the location of the central meridian. Ambiguous symbolization schemes and incompatible or vague feature definitions might require undue inference or guesswork. Maps based on photogrammetric surveys are generally more accurate than maps based on field reconnaissance or plane-table surveys, and a survey using the same *datum*, or model of the earth's shape, as the compilation base should have fewer discrepancies than a map based on an earlier datum or, worse yet, no higher-order geodetic foundation whatever.[53] Enlarging or reducing one source to the same approximate scale as another and overlaying the two maps on a light box can provide useful insights into their relative reliability. Because some map publishers or surveyors have been more conscientious or used better methods than others, the writings of historians of cartography can help the map author evaluate competing sources.[54]

Dates on maps require cautious interpretation. When using modern sources, the map author should read the fine print. For example, the prominent 1990 publication date on a Geological Survey topographic map might promise greater accuracy than the statement in smaller type that the map is based on air photos taken in 1987 but "field checked" (in a limited way, to resolve ambiguities in photo interpretation) in 1989. Moreover, although a note on a 1:100,000-scale map dated 1990 says "compiled from 1:25,000-scale topographic maps dated 1987–88," further investigation might reveal that these earlier maps were based on still earlier photography, from 1985. Unlike Geological Survey maps, most commercial products rarely describe their origin. A publisher might simply substitute a current date for an earlier one to make the product more competitive, or compile a map from earlier sources without incorporating changes in boundaries, names, or other information.

Avoiding perpetuation of old errors sometimes requires a critical examination of how well a map reflects the time it purports to portray. Scholars using medieval maps should be aware that sources with a vague date, such as "ca. 1275," might cover several decades, or even a century or more. The famous world map in Britain's Hereford Cathedral, for instance, reflects numerous redrawings during the twelfth and thirteenth centuries and is not a valid geographic snapshot for any single year.[55] Potentially more problematic are modern cartographic databases, which can reflect to-the-minute changes for some features yet perpetuate erroneous or obsolete representations for others.[56]

Scholars using pre-twentieth-century cartographic sources must watch out for inconsistent meridians.[57] Universal acceptance of the Greenwich meridian as the prime meridian was not complete until after World War I, and at earlier times various national surveys measured longitude according to a prime meridian through Cadiz, Naples, or Paris.

Careful assessment of source materials can help the map author develop an appropriate order of compilation. The sequence in which the cartographer transfers features onto the compilation base is important, because the need to avoid graphic interference requires that features added later are displaced more than those transferred earlier. The map's goals and scale of concepts should be helpful in deciding which features warrant more geometric accuracy and which can tolerate greater cartographic license.

Among the many pitfalls of multisource compilation, one of the more serious is incompatible projections. When projections are identical, the map author can use a copy machine or computer graphics system to enlarge or reduce the source image to the approximate scale of the compilation base, and then can simply overlay the two maps and copy features onto the compilation base. But when incompatible projections make direct retracing impossible, corresponding networks of equivalent features are needed to guide the transfer of points and lines from one projection to another.

Figure 5.10 illustrates how a dense grid of meridians and parallels or shared landmark features promotes visual transfer. For instance, in transferring the railway from the source map to the compilation, the map author should note that it crosses the river at Center Robinson and runs straight through the city of Snyder. Relative position also guides the visual transfer of a town lying in the triangle formed by three nearby cities. Although sophisticated mathematical transformations and computer algorithms can blend images or adjust one projection to another, most map authors using scanners and electronic graphics systems still "eyeball it" when transferring and redrawing features.[58]

A reliable compilation base helps the map maker control error. Because a cartographic feature tends to lose accuracy with every transfer from one map to another, a map that has evolved through several generations of compilation may have accumulated considerable distortion. Using a map from a reputable map publisher or a government agency as the compilation base is a good way to compensate for other source materials of uncertain or questionable reliability. Cartographers must also be wary of shrinkage or swelling of paper maps, lens and emulsion distortion in photography, and stretching or shrinking in one direction by a copy machine or electronic scanner.[59]

Scholars must also be careful when compiling maps from aerial photographs, which can severely distort distance and shape. Air photos show many landscape details missing from large-scale topographic maps and are useful cartographic sources for anthropologists, industrial archaeologists, military historians, rural sociologists, and many other social scientists. But as Figure 5.11 illustrates, the aerial photograph is a perspective view in which lines of sight converge at the camera's lens, and vertical cliffs, flag poles, and the faces of tall buildings often appear to be tilted or lying on their sides. The geometry of the air photo displaces the top of a vertical feature outward from its base along a radius from the center of the photo, and this displacement can be enormous in mountainous, "high relief" areas. Cartographers call this effect "radial displacement due to relief." Especially pronounced for tall features or rough terrain toward the periphery of the photo, radial displacement also distorts distances between features. When compiling a map by copying points and lines from air photos, the careful map author avoids direct tracing and

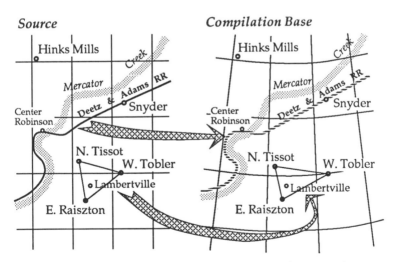

FIGURE 5.10. Meridians, parallels, and similar landmarks support the visual, "eyeball" transfer of features from a cartographic source (left) to the compilation base (right).

never transfers features near the edge of the photo unless the photo and the compilation base share a dense network of similar features.[60]

If each cartographic source covers only part of the area, sections of roads, railways, boundaries, rivers, and other continuous linear features transferred from different sources often don't meet and blend smoothly. Misalignment even occurs with government quadrangle maps, especially if adjoining sheets differ in scale, survey method, or publication date. Where features that should meet don't, the map author must decide which source most likely represents the correct alignment and carry the trend of the feature from the more accurate source into the area covered by the less accurate source. Although edge matching can be highly subjective, blending features to make them look natural is almost always preferable to showing awkward and unrealistic right-angle jogs or failing to connect continuous features.

Cartographers sometimes compile maps by merging or integrating computer files of boundaries or point locations. When the data have been "captured" from paper maps with a digitizer

Profile

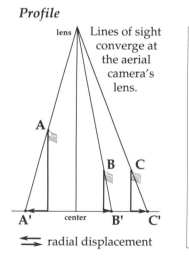

lens   Lines of sight converge at the aerial camera's lens.

A

B   C

A'   center   B'   C'

⇆ radial displacement

Aerial Photo

Although flagpoles B and C are equally tall, displacement is greater for the top of C, which is farther from the center.

A'   center   B'   C'

Bases of flagpoles A and B are equal distances from the photo's center, but top of taller flagpole A is displaced farther outward from its base than top of flagpole B.

FIGURE 5.11. Because the aerial photograph is a perspective view (left), the top and bottom of a tall, perfectly vertical feature lie at different points along a line radiating from the center of the photo. The effect of radial displacement increases toward the edge of the photo, as well as with the height of the object (right).

or scanner, in most cases further manipulation is needed to remove graphic interference, resolve discrepancies among sources, and blend features along edges. But computer-based compilation can be straightforward if all coordinates in the data measure latitude-longitude or reflect the same projection or grid system. For example, plotting point locations and a file of state boundaries on the same projection can be an easy way for a political scientist to make a revealing map of the large U.S. cities with Republican mayors. Scholars using electronic data should explore the potential benefits of map-making software and readily available, inexpensive boundary and point-location files.[61]

Reproducing Facsimiles

Facsimiles of significant or representative maps are useful both as documentary evidence to prove a point and as decorative art to catch the reader's attention and add interest to otherwise dry material.[62] But the author must be clear about the illustration's role in the book or article, or the format chosen for the cartographic facsimile may fail, or even repel the reader. For instance, a facsimile intended to show the reader how part of a city or town looked on an 1897 map could be far more effective as a full-size excerpt than as a halftone photograph that would reduce the whole map to a meaningless hodgepodge of illegible type and fuzzy or coalescing lines. If the purpose of a black-and-white facsimile is to point out a small feature or place name on a complex map originally printed in a dozen colors, the author needs to enlarge the significant part of the map and perhaps highlight particularly relevant symbols or labels. In contrast, if a feature such as a railway or coastline extends across a large map, the best strategy might be to use two illustrations: a reduced, redrawn and carefully generalized version of the map focusing on the important feature; and a full-size or enlarged facsimile excerpt to show symbols, type, and detail. Effective use of a cartographic facsimile involves far more than reducing the whole map to a page-size illustration.

A good way to begin is to determine why you want to show the reader how a map looks and then to draft the words to tell why the map is noteworthy and what it shows. As you write, look at the map and also make a rough sketch to help visualize the enlarged, reduced, or full-size image you will give the reader. If you are uncertain about the sketch, find a copy machine that enlarges and reduces, cut out page-size or smaller portions of the map image, and experiment with tradeoffs between detail and coverage. Treat this sketch or photocopy as a crucial part of writing your draft. And don't consider the draft complete until you have not only written what you need to say but also are certain that the reader can see what you want to be seen. Although it is possible for an author to add facsimile maps to an otherwise complete and polished manuscript, insertions at the later stages of writing require considerable effort in order to integrate new information, promote cohesion, and maintain a smooth, logical discussion.

The quality of the map that appears in your book or article depends on what you give the printer. If a color map is to be reproduced as a black-and-white halftone and you can send the original, do so. But be specific about which features are important and insist on seeing a proof. It is good to consult with your publisher's art director, because some color combinations, such as red lines overprinted on black symbols, are difficult and costly to resolve. If the original map is in a university library's special collections department, you might arrange for the university's graphic arts unit to make a reproduction negative, from which you or your publisher can then make prints and halftones. The Library of Congress and the British Library have their own photoreproduction units, which can provide reproduction negatives for a nominal charge. If the copy is a reduction, be certain to keep a record of the map's original dimensions. If the map appeared in a book, newspaper, or magazine, find out if the author, artist, or publisher who granted permission to copy it might also be able to supply a high-quality print or the original artwork.

Many straightforward maps, including most topographic maps, can be reproduced as line art, which avoids breaking the image into the grid of tiny dots used for halftones. Maps printed as line art offer a relatively sharp image for the reader and fewer headaches for the author and the publisher. In general, if the map has few color tints or is itself a black-and-white line illustration, consider line reproduction, not halftone.

When ordering copies from a map collection's photoreproduction unit, request a high-contrast paper print, not a continuous-tone reproduction negative. The print is less expensive, shows you how the facsimile will look, and even allows for some minor retouching. A copy machine that is regularly cleaned and resupplied with toner can also make suitable copies of line art; the process is inexpensive, requires only a short wait, and can encourage experimentation with different exposures. The copier also promotes the experimentation needed if shorter exposures drop blue lines representing streams or canals and longer exposures pick up oxidation stains on aging paper. A copy machine that can change scale might also allow you to judge how the map would look if reduced to fit the page or enlarged to highlight particular symbols.

Line illustrations, such as maps copied from books and newspapers, can be enlarged photographically, touched up, and then reduced down to reproduction size. As an example, for Figure 5.6 I restored a map originally engraved on a wood block and printed in a newspaper. I brought the image back from the Library of Congress as a photocopy, enlarged it to about 250 percent as a high-contrast paper print, removed stray specks of ink or toner with correction fluid and an X-acto knife, touched up fuzzy lines and imperfectly inked type with a fine-tip red marking pen, and reduced the restored image to a size smaller than the original. Although news photographers and printing historians might frown at the practice, restoration of an image degraded by imprecise printing and acid paper compensates in part for the reduction necessary to fit it into a book or academic journal. Restoration is particularly helpful when a cartographic image must be retrieved from microform as a less-than-perfect machine copy.

Not all line illustrations warrant retouching. Figure 5.3 was shot with high-contrast photographic paper from a late-nineteenth-century topographic map printed in black, brown (for contour lines), and blue (for water features). Several test prints were needed to find an exposure that captured the map's blue lines without blurring its black type. Although not perfect, the result is both informatively detailed and aesthetically acceptable. Because of the map's dense symbols and small labels, attempting a restoration would have been risky as well as tedious.

When presenting a facsimile map as a text or as visual evidence, the scholar must consider how reproduction might strengthen or weaken the visual impact of the original. An accurate description of the original can be important, for example, in a linguistics or political science work using a map as a semiotic artifact in a study of propaganda maps. After all, when an author treats a map as a text, the cartographic facsimile becomes a graphic quote. Because size, color, and location within a larger publication can be important as parts of a map's context, offering an inexact and misleading facsimile without explanation is as unethical as quoting a source out of context.

In Summary: Seek and Collect

If this chapter has been at all effective, you should have a fuller understanding of why the scholar in the early stages of research needs to think about using maps. You should also want to visit the map collection of a good research library soon, or, if you have already introduced yourself to a strong map collection, to visit another—the Library of Congress, the Newberry Library in Chicago, the British Library, or the Bibliothèque Nationale in Paris, for instance. Or it may be more relevant to your interests to visit a geographically focused map collection at a city historical society or at state or provincial archives. If it has been at all effective, this chapter has given you a clearer sense of what to look for and what questions to ask.

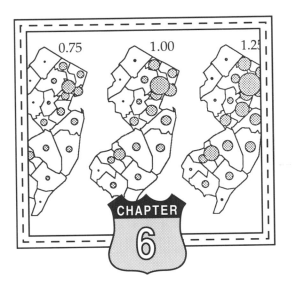

Statistical Maps,
Data Scaling,
and Data
Classification

QUANTITATIVE DATA AGGREGATED BY CENSUS TRACT, ward, county, state, province, or country support empirical investigations in many branches of social science. Studies of commuting behavior, land value, urban change, and other phenomena affected by distance need these data to detect spatial pattern, and studies of housing, electoral behavior, and poverty rely upon data aggregated by area for information about income and ethnicity. Social scientists need maps to explore and understand their data and to confirm and refine their hypotheses. They also need to understand the principles and limitations of statistical maps, which are not as straightforward as many software products suggest. This chapter examines the pitfalls of mapping software, explains the need for compatibility between the map and the data, and presents a strategy for the effective design of data maps.

Two complementary types of statistical map apply graphic logic to meet most needs in social science: graduated-point-symbol maps use vertical bars or circles to show variation in magnitude, and choropleth maps use graytones or carefully graduated colors to show variation in intensity. This chapter focuses on these two basic types, their inherent complementarity, and methods for scaling and classifying the data. Separate sections on mapping count data and mapping intensity data draw heavily on the discussion of the logic of graphic symbols in chapter 3 and the examination of communication goals and the design of map titles and keys in chapter 4. Useful modifications of these basic strategies for mapping quantitative data are also explored.

Mapping Count Data

Count data and intensity data are different. Census tabulations and other numerical databases commonly report areally aggregated data as counts, such as the population of a census tract or the number of infant deaths by county of residence. As I explained in chapter 2, these counts are magnitude data, and their effective cartographic portrayal requires a graduated-point-symbol map, so that size is the principal visual variable. In contrast, area symbols that vary in value (graytone on a choropleth map) are appropriate for mapping population density, median income, infant mortality rate, and other intensity measures. The choropleth map that displays raw count data tends to confuse or mislead the viewer.

The software industry has largely ignored the fundamental difference between count data and intensity data. As a consequence, the social scientist who gives a computer mapping program count data almost always gets a choropleth map, in which the principal visual variable is value, not size. This incompatibility between map and data reflects the industry's "success experience" marketing tactic, in which "user friendly" software conjures up a map without pestering the naive user to describe the data or specify a particular type of symbol. With most mapping software, the choropleth map is the default display that results when the user does not deliberately invoke certain options. Software developers apparently think the choropleth map is more impressive or more predictable than the graduated-point-symbol map, even though count data are more common than intensity data.

Figure 6.1 illustrates how a choropleth display of magnitude data can mislead the viewer by exaggerating the relative importance of one part of the region. The variable mapped here is the number of infant deaths during 1988 for New Jersey's twenty-one counties. An otherwise competent mapping program produced this map the instant I finished entering the data.[1] Note the single county in the highest category, represented on the map by the darkest symbol: this is Essex County, which contains New Jersey's largest city, Newark. Because the second highest category is empty, Essex County stands out from the other counties. Yet less than a fifth of the state's infant deaths occurred within the boundaries of Essex County.

Does the use of county units to aggregate the data in Figure 6.1 explain the spurious prominence given Essex County? To address this question we need to consider a hypothetical redrawing of county boundaries. Note the four smaller counties adjacent to Essex County on the north, east, and south; although these four counties collectively had more infant deaths than Essex County, each individually is only in the third or the fourth category. Hypothetically, these three areas could be combined into a single political unit that would surpass Essex in the infant mortality count and would be more visually prominent on a new map, because of its larger size and its darker symbol. A second hypothetical readjustment of boundaries could divide Essex County into four smaller units with counts and symbols similar to Essex's neighbors on the current map. The point of making these hypothetical reapportionments is to demonstrate that the number, shape, and areal extent of the data areas affect both whether and where the viewer will see a peak area. The choropleth map in Figure 6.1 is deceptive because it represents infant death as a problem strongly focused on Essex County and exaggerates the relative importance of this single political unit.

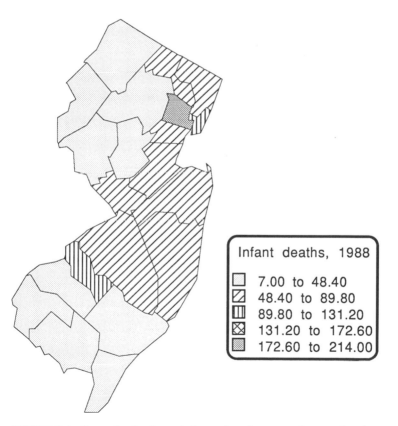

FIGURE 6.1. Example of a choropleth map based on count data, produced as a mapping software package's default display.

To understand fully why this default map is misleading, you also need to examine Figure 6.2, a graduated-point-symbol map that offers a more graphically logical view of same data. Making this map required not only a sense of how a map should display count data but also a working knowledge of the software and a little effort.[2] But the result is a demonstrably better map, which uses big circles to point out counties with many infant deaths and tiny circles to indicate counties with few infant deaths. The viewer does not have to look at the key to determine which counties had more infant deaths, and the map doesn't lump diverse counts such as "7.00 to 48.40" into a single uninformative category. Not, too, that Essex County emerges not as a relatively unique trouble spot but as part of a cluster of counties with substantial numbers of infant deaths. Even the hypothetical merging and subdividing of counties suggested in the preceding paragraph would not destroy the stable geographic pattern of infant deaths portrayed in Figure 6.2.

The design and content of Figure 6.2 helps the viewer in several other ways. To promote value estimation, I moved to the front

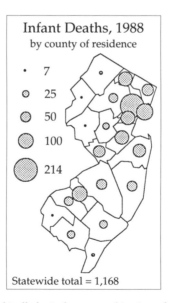

FIGURE 6.2. A graphically logical cartographic view of the count data used in Figure 6.1.

two circles otherwise partly blocked by the large circle for Essex County. The map's concise title includes a year-date, and a subtitle notes that the data have been aggregated by county of residence, not county of occurrence. Its key uses sample circles, or *anchor stimuli*, for the minimum and maximum county-unit values to show the range of the data. The key also promotes accurate value estimation by including several rounded yet representative intermediate anchor values. A statewide total at the base of the map offers a fuller sense of the seriousness of infant death as a health problem in New Jersey and helps the viewer put individual county-unit and anchor-stimulus values in perspective.

As Figure 6.2 demonstrates, a map's foreground layer of magnitude symbols should stand out from the area boundaries in the background. Figure 6.3 shows several other acceptable designs, all with the proportional-circle symbols positioned in front of a network of relatively thin boundary lines. Although using outlined circles with transparent interiors might promote clarity on maps with many overlapping circles, the effectiveness of outlined circles depends on a marked contrast between the circles and the boundary lines. Solid black circles draw the eye more strongly to

Foreground-background contrast for graduated-point symbols

Outlined circles   Tinted circles   Solid circles

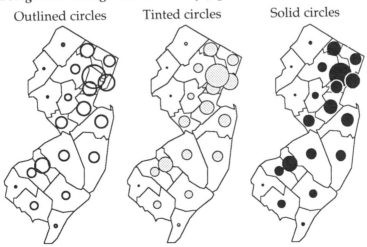

FIGURE 6.3. Three approaches to providing contrast between graduated point symbols and the map's geographic frame of reference.

clusters and concentrations but can give the map an ugly, harsh look. Tinted circles offer a useful compromise between overly harsh solid circles and graphically weak outlined circles. Experimentation with different designs can help the map author achieve necessary contrast, appropriate emphasis, and even aesthetic elegance.

A second important design decision is the data-scaling relationship, which specifies the circle size for each count portrayed on the map and in the key. Experimentation with data scaling is essential to avoid either an unrevealing, inexpressive pattern of symbols that are too small, or a cluttered, confusing pattern of symbols that are too large. Mapping software usually generates circular point symbols proportional in area to the data values they represent, so that a symbol representing two hundred deaths will be twice as large as a symbol representing one hundred deaths. The user controls data scaling by specifying the diameter of the largest circle on the map.

Figure 6.4 illustrates the visual effects of four maximum diameters I considered in designing Figure 6.2. The number above each example is the ratio of its maximum diameter to the maximum diameter used for Figure 6.2, which is represented in Figure 6.4 by the second example from the left. The leftmost example in Figure 6.4, with a ratio of 0.75, indicating a maximum diameter 25 percent smaller, yields a comparatively weak pattern, in which the Essex County circle barely touches its two closest neighbors. This map not only offers a less sensitive representation of differences among the counties but also plays down the concentration of infant deaths in and around Essex County. In contrast, the example at the far right, with a ratio of 1.50, indicating a maximum diameter 50 percent larger than in Figure 6.2, has substantial overlap, especially for the Essex County circle, which covers most of two neighboring circles. Moving these hidden circles to the front would improve value estimation, but the map would still be needlessly cluttered.[3] Overlap is slightly less severe in the second example from the right, which has a ratio of 1.25. I chose the smaller maximum diameter of the second example from the left, in which the circumference of the Essex County circle penetrates no farther than the center of any overlapping circle. This criterion is a useful rule of thumb if it is not used slavishly.

Figure 6.4 illustrates the inherent subjectivity of graduated-point-symbol maps. The puny circles produced by a small maximum diameter might seem to play down the seriousness of infant mortality, whereas the enormous circles produced by a large maximum diameter seem to exaggerate its prevalence by covering a large part of the mapped area. But for a pair of maps that invites visual comparison, puny circles on one map and crowded circles on the other may be required for the sake of probity. For instance, maps comparing infant mortality for 1908 and 1988 should employ the same data-function scaling in order to reflect accurately the enormous advances in health care. And a pair of maps comparing infant mortality for New Jersey's two major racial groups should use identical scaling to indicate clearly and perhaps dramatically the disparity between statistics for the state's African-American and white populations. Forcing mapping software to scale both sets of data the same way might call for some trickery, such as adding an extra data point with the same large value to both maps and eliminating this extra symbol after the maps have been produced.

To avoid exaggerating the importance of maps of count data, I do not discuss here additional refinements of graduated-point-symbol maps. Several other symbolization strategies appropriate for count data have already been mentioned in chapter 3, including one-dimensional vertical or horizontal bars, squares or picto-

Relative diameter of the largest graduated circle

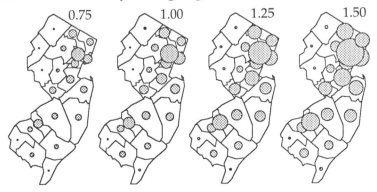

FIGURE 6.4. Different scaling factors, yielding graduated circles that differ in degree of overlap. The scaling factor is the diameter of the maximum circle, relative to the maximum diameter used for Figure 6.2.

rial point symbols in two dimensions, cubes and spheres in three dimensions, and apparent-value rescaling to correct for the visual underestimation of the size of larger symbols. Commercial mapping software supports few of these strategies, and but for most purposes graduated circles will be quite adequate, if the map author experiments with data scaling and foreground-background contrast, as examined here. Proportional-point-symbol maps are quite robust for portraying counts aggregated by areal units, but the map author should not expect the map to do much more than show which parts of a region have lots of the phenomenon in question and which have little.

In short, a graduated-point-symbol map used alone is seldom meaningful. For example, a reader only vaguely aware of the geography of the northeastern United States probably would not find my graduated-circle map of infant deaths in New Jersey (Figure 6.2) particularly informative, because its pattern of large

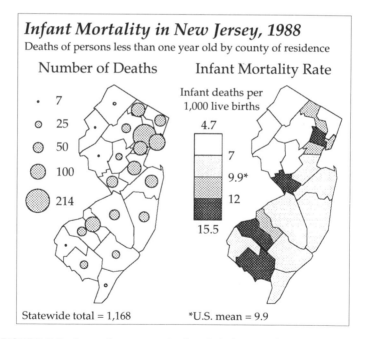

FIGURE 6.5. A complementary pair of statistical maps using proportional point symbols (left) for count data and a choropleth map (right) for meaningful rates based on these counts.

and small circles largely reflects population distribution. That Essex County has the largest circle is neither surprising nor revealing because Essex County has more people than any of the state's other twenty counties. The counties with very small circles have relatively few people, relatively few births, and relatively few infant deaths. The map might be useful for showing where an infant-care research center should be located or where funds for prenatal clinics should be allocated, but not for much more.

To make a map of count data useful, the map author usually needs to combine it with a choropleth map of intensity data. Figure 6.5 illustrates this point for the infant mortality statistics from New Jersey. The right-hand map is based on a comparison of the number of infant deaths to the number of live births. Its graytone symbols show that the infant death rate is comparatively high not only in Essex County but also in Mercer County (in the center of the state, around Trenton) and in two counties in the south. But a quick check of the graduated-circle map to the left reveals that the two high-rate counties in southern New Jersey had fewer than fifty infant deaths each. Moreover, the area surrounding Essex County is far from uniform on the rate map; counties directly north, east, and south of Essex are in the second highest category, and still above the national average. But Bergen County, in New Jersey's northeast corner, is in the lowest category, despite having over fifty infant deaths. Very generally, though, the state's less populated counties in the northwest and southeast tended to have lower-than-average rates and comparatively few infant deaths. Although the mortality rate map is clearly the more meaningful of the two maps, the graduated-circle map shows marked geographic variation in numbers of infant deaths and helps the viewer interpret the map of rates. The pair of maps succeeds because it makes the viewer want to examine other factors, such as race, affluence, and health care.

MAPPING INTENSITY DATA

Intensity data usually are more informative than count data, but designing an informative choropleth map can be more complicated than designing an effective graduated-circle map. The map author faces five important problems: how many categories to use; how to make these categories reflect significant trends in the data; how to show progressive increases in intensity with an unambiguous series of graphically stable area symbols; how to describe the intensity variable clearly and concisely; and how to link the symbols, classification, and intensity measurements with an informative, easily interpreted map key. Although guidelines such as those in this section help the social scientist narrow the range of choices, experimenting with categories and symbolization is necessary in order to recognize trends in the data and determine how to let those trends speak to the viewer through the map.

Informative Classifications

Because choropleth maps require more decisions than graduated-circle maps, the naive map author is more easily seduced by uninformative default choropleth solutions proposed by mapping software. Figure 6.6, the default map my computer mapping program proposed for the infant-death rate statistics mapped in right half of Figure 6.5, is a good example of the mindless default display the careful scholar must learn to look beyond. It has several deficiencies: a key with frustratingly small area-symbol boxes, a series of graytones that mixes dot screens and line screens, and an arbitrary equal-interval classification scheme that divides the range of the data (from 4.71 to 15.48) into five equal chunks. Comparatively coarse line patterns, one with crossed lines, separate fine-dot area symbols at the upper and lower ends of the gray scale. Slight overinking during printing might make the symbol for the lower category darker than one of the line patterns, or a slight overexposure during plate making could make the dot pattern disappear. To keep straight which of two area symbols represents higher values, the conscientious viewer must refer frequently to the cryptic samples in the stingy key. The default classification ignores the inherently meaningful national mean (9.9), which falls near the middle of the third class (9.02–11.17). That essentially the same general pattern is found in the right half of Figure 6.5 is partly luck and partly a reflection of the robustness of choropleth maps.

Little more is said here about graytone area symbols than to warn of the danger of violating the graphic logic of area symbols (presented in detail in chapter 3), or of attempting to reproduce symbols consisting of densely packed grids of fine dots, which a printer may inadvertently darken or lighten. Overinking during printing often is a consequence of a map author's eagerness to have at least five categories, but possibly six or even seven. There must be something mystical about five, the number of default categories used by every choropleth mapping program I've ever seen. Yet it is better to represent fewer categories with graphically stable area symbols than to risk a misleading, unreliable, or aesthetically gauche map.

Using only a few categories need not mean producing an uninformative map. Carefully chosen breaks between categories allow the viewer to identify areas above or below a meaningful value, such as a national, provincial, or state average rate. As an example, in the choropleth map in Figure 6.5 a category break at 9.9 allows the classification to point out the eight New Jersey counties with infant mortality rates higher than the nationwide average. If the state rate had been substantially different from the

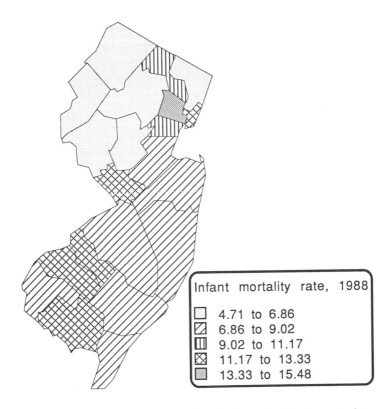

FIGURE 6.6. Example of a default choropleth map based on intensity data.

national rate, the map could have had two meaningful breaks. Choropleth maps depicting a rate of change might also have two meaningful breaks: 0.0, to distinguish places that gained from those that lost, and the regionwide rate of change, to differentiate places changing comparatively rapidly from those changing more slowly.

Breaks pointing out significantly high or low data values are especially useful. When author and audience share a background in basic statistics, the mean μ and standard deviation s provide several meaningful breaks. For example, the four breaks at $\mu - 1.5s$, $\mu - 0.5s$, $\mu + 0.5s$, and $\mu + 1.5s$ yield five categories that can be described as extremely low, low, average, high, and extremely high. Less numerically astute readers confronted with a choropleth map showing many areas might appreciate knowing which areas are among the highest or lowest quarter, fifth, tenth, or twentieth. Of course, the key or title must indicate why a break, category, or classing scheme is meaningful. In Figure 6.5, for example, the importance of the break at 9.9 infant deaths per 1,000 would be lost without the prominent footnote.

Users of mapping software should be wary of equal-interval and other easily programmed classing schemes, as illustrated in Figure 6.7. These widely used approaches are in no way standard, in the sense of being intrinsically revealing, universally appropriate, or even officially endorsed. The equal-interval strategy has the appeal of simplicity, but little more; a programmer needs only to find the distribution's highest and lowest values and then to partition the range equally among the categories. Yet when a few values are very high, the three middle categories of a five-category map could be empty. Another widely used classing scheme, the *quantile* approach, is only slightly more computationally complex; the programmer must rank-order the data values from lowest to highest and then allocate an equal or nearly equal number of places to each category.[4] When confronted by many low and a very few high data values, a five-category quantile scheme might yield four lower "quintiles" with very narrow ranges and an upper quintile with a broad range, including some very high and a few rather low values. Perhaps the best use of these simple classifications is to generate multiple views rapidly, so that the map author can see how sensitive the map is to different classing techniques and different numbers of categories.

An interactive yet systematic approach to designing choropleth maps might begin with a simple *number-line plot*, which represents data values as points scaled along a parallel arithmetic axis, as in Figure 6.8. Indeed, if mapping software must use a default-display approach, perhaps it could best begin with a number-line plot, asking the user first to enter the regionwide average or another meaningful break and then to select breaks that respect natural clusters as revealed by the plot. Breaking up a cluster of highly similar data values by putting an arbitrary

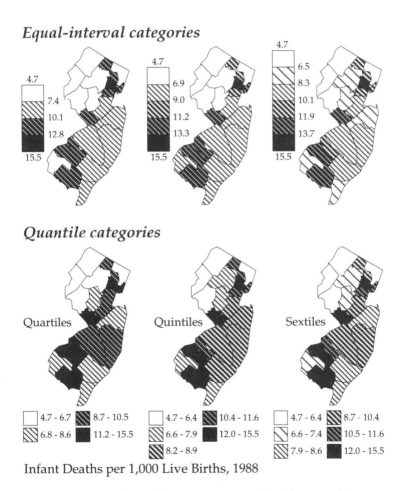

FIGURE 6.7. Four-, five-, and six-category maps of the infant mortality rate based on equal-interval and quantile classing schemes.

category break in the middle seems unwise, especially if relatively homogeneous categories might be used instead. Yet the map author who selects an equal-interval, quantile, or other rigid classing strategy without first looking at the data might do just that. A number-line plot may point out an optimum number of categories or reveal a uniform distribution, for which quantile or equal-interval classifications would yield similar and generally appropriate results.

Number-line plots can also help the map author cope with outliers, that is, extremely high or extremely low data values at the outer end of the data range, well removed from the majority of values, as shown in the upper plot in Figure 6.9. When outli-

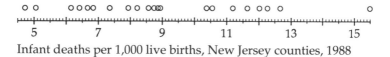

5 7 9 11 13 15
Infant deaths per 1,000 live births, New Jersey counties, 1988

FIGURE 6.8. A number-line plot for the New Jersey infant mortality data.

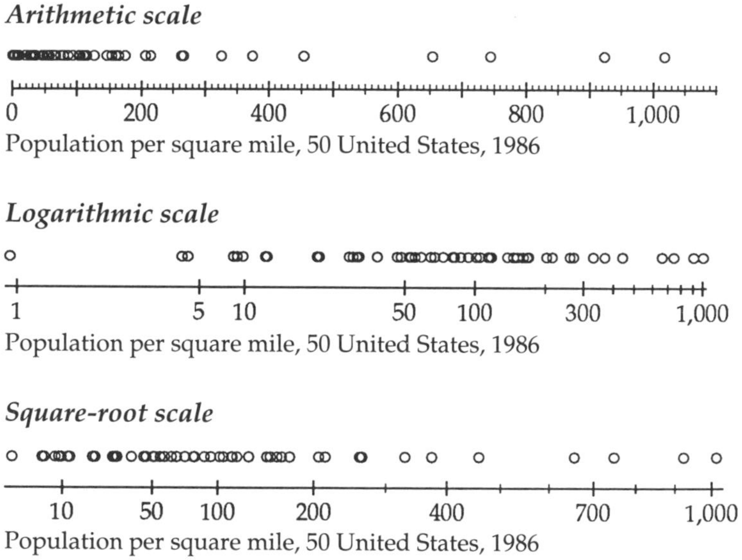

Arithmetic scale
0 200 400 600 800 1,000
Population per square mile, 50 United States, 1986
Logarithmic scale
1 5 10 50 100 300 1,000
Population per square mile, 50 United States, 1986
Square-root scale
10 50 100 200 400 700 1,000
Population per square mile, 50 United States, 1986

FIGURE 6.9. Number-line plots for a skewed distribution represented using arithmetic, logarithmic, and square-root scales.

ers arise, as they often do if areal units vary widely in area or population, knowing the quality of the data is essential. If outliers obviously reflect errors in measurement, the best graphic strategy might be to use a blank symbol (often reserved for missing data) with an asterisk in the middle pointing to an explanation at the bottom of the map. If the data include a few very large but highly dissimilar values, grouping them into a single category can be appropriate, especially if the author must discuss each individually. When addressing a quantitatively savvy audience familiar with scale transformations, the social scientist might plot the data on a logarithmic or square-root scale, which usually yields a less skewed, more uniform distribution, as the lower two plots in Figure 6.9 demonstrate. The map key can use either transformed or nontransformed numerical labels, or perhaps both; a subtitle or note in the key should mention either the actual transformation of data values or the use of a transformation in setting category breaks.

For choropleth maps with a very large number of areal units, a number-line plot might not work, because the small point symbols representing data values would be too numerous and too cluttered. In such cases, a frequency histogram is a necessary and acceptable substitute. But to provide graphic precision comparable to that of a number-line plot, the histogram must have at least twenty categories.

Informative Keys

The map key, which links the choropleth map's data values, classification, and graded series of graytone area symbols, deserves careful thought. In addition to describing the area symbol and the range of data values corresponding to each category, the key can describe each category's size, internal homogeneity, and relative share of the overall range of data values. The map key can also link the categories to the number-line plot or frequency histogram and can relate the area symbols on the map to meaningful breaks or categories. For some applications the key might even show the population or land area for each category, to help the viewer evaluate the overall classification and the relative importance of individual categories.

Although not every choropleth map needs a rich and complex key, a few examples illustrate how the map author may creatively and effectively adapt these concepts. Figure 6.10 links the number-line plot of infant mortality rates to the four categories of the New Jersey choropleth map in Figure 6.5. Two versions of this key illustrate gapped scales and continuous scales. Gapped-scale labels and graytone samples show the actual range of each category, whereas continuous-scale labels and graytone samples focus on the overall classification, rather than on its individual categories. A gapped-scale key would call attention to significant natural breaks in the distribution of data values, whereas a continuous-scale key would more readily reflect the rationale or strategy of a classification based on equal intervals, the mean and standard deviation, or a logarithmic or square-root transformation. Figure 6.11, which presents vertical versions of both designs, also demonstrates a further graphic transition to equal-size graytone samples. This addition is especially useful when one or more categories occupy a very narrow portion of the overall range.

Figure 6.12 illustrates a key with a frequency histogram showing the relative population of each of the four categories in Figure 6.5. Note the left-to-right progression from (1) the data

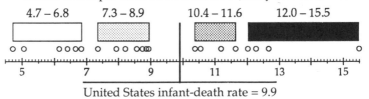

Continuous-scale labeling

Infant deaths per 1,000 live births, New Jersey counties, 1988

United States infant-death rate = 9.9

Gapped-scale labeling

Infant deaths per 1,000 live births, New Jersey counties, 1988

United States infant-death rate = 9.9

FIGURE 6.10. Examples of continuous-scale and gapped-scaling labeling for a horizontal key.

scale and distribution of individual data values, with a meaningful break based on the nationwide mean, to (2) graytone samples representing each category's relative range, (3) equal-size graytone samples with gapped-scale labeling, and (4) graytone samples embedded in histogram bars reflecting each category's relative importance. Because the histogram in Figure 6.12 shows only moderate variation in population size, the reader may infer that the categories are more or less equally significant. A histogram would be especially revealing if areas with small populations differ radically from areas with large populations.

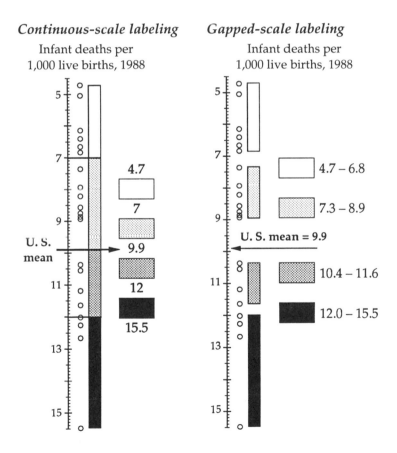

FIGURE 6.11. Examples of continuous-scale and gapped-scale labeling for a vertical key.

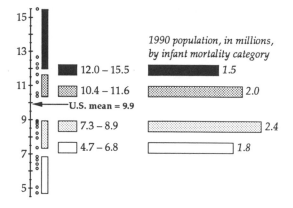

FIGURE 6.12. Example of vertical key with a histogram describing the distribution of population among categories.

The map key in Figure 6.12 differs from other map keys in this chapter because its categories increase in value toward the top, not toward the bottom. Statisticians tend to favor this upward-increasing strategy, modeled after the vertical axes of bar charts and scatterplots. Yet social scientists seem to prefer the downward-increasing strategy used to organize rows in census reports, cross-tabulations, and data tables in general. Although either approach is valid, consistency within a book, article, or series of maps will avoid confusing the reader. A useful compromise is the horizontal key, as in Figure 6.10, for which both the statistical-axis metaphor and the table-row metaphor place higher categories farther to the right.

What Are Intensity Data?

Misuse of choropleth maps can occur when the map author lacks a clear sense of what makes a variable an intensity measure. Because most demographic and economic data appropriate for choropleth mapping are based on enumerations or surveys, the social scientist must make certain that raw counts have been implicitly or explicitly adjusted for variation in land area, number of residents, or some other relevant magnitude. A good example is the number of persons residing in places considered urban. The raw count is not satisfactory, because counties with more people might be expected to have more urban residents. But dividing the urban population by the total population yields the urban percentage of the population, an acceptably adjusted intensity measure. Other representative examples of intensity measures are population density (persons per square mile or square kilometer), mean household income, median household income, taxes paid per capita, number of persons per room, percentage of households receiving public assistance, and population change between 1980 and 1990 expressed as a percentage of the 1980 population. In general, choropleth maps may be based on densities, means, medians, rates, ratios, percentages, and percentage rates.

Because magnitude measures sometimes masquerade as percentages, be especially wary of the data table with a column labeled "percent" next to a column labeled "number." For our New Jersey example, a table of county-unit data might show a column of infant death counts next to a "percent" column that was computed merely by dividing the count for each county by the total for the state. As a result, the "percent" column sums to 100, is perfectly correlated with the "number" column, and can yield a choropleth map as conceptually flawed as Figure 6.1, the misleading default choropleth map deconstructed at the beginning of this chapter. This example of spurious "percent" data demonstrates that merely remembering the densities-rates-percentages rule is no substitute for truly understanding why maps based on intensity symbols require intensity data.

Modifications for Greater Effectiveness

Effective communication with statistical maps often requires small but significant modifications of the data or the display. This section examines a number of further adjustments the map author might consider.

Visible Symbols

An important concern, often overlooked by users of mapping software, is that the viewer should be able to see all the symbols on the map. On a graduated-circle map, for instance, if large symbols would hide or grossly overshadow small ones, the map author might want to place the smaller of two overlapping circles in front (or on top) of the larger one, move the symbols apart slightly to reduce overlap, or provide a detail inset map showing the congested area at a larger scale. Detail insets are also useful on choropleth maps, especially if the areal units vary greatly in size.

Maps of New York City can provide good illustrations of the three strategies for designing simple detail insets. On a standard county-unit choropleth map of New York State, the five counties that make up New York City often are too small for the viewer to readily discern their area symbols. A detail inset is particularly appropriate here, because about half the state's population, and much of its political clout, resides within the five "boroughs." Figure 6.13 shows three insets differing in level of generalization. The example on the left offers a comparatively precise representation of boundaries and relative size, but doesn't quite provide an adequate "graphic platform'" for the area symbol representing New York County, otherwise known as Manhattan and surely the borough with the most clout. The design in the center is a polygon representation similar in detail to the county outlines used by a number of mapping software packages; the width of Manhattan is exaggerated to make its area symbol easier to discern. The inset at the right treats all five counties as equal-size boxes identified by name and relative position; viewers not acquainted with the political geography of New York would find this approach helpful if labels link "Manhattan" and other well-known borough names with their corresponding county names. On a New York State map that includes an inset, a single areal unit on the main map might usefully represent a five-county average value for New York City as a whole.

The "visibility base map" in Figure 6.14 extends the exaggerated-width and polygon-representation concepts to a state-unit map of the entire United States. On a small choropleth map with a conventional projection, viewers often cannot easily decode the small graytone area symbols for Delaware and Rhode Island. But distortions in the base map in Figure 6.14 make the symbols of these small states visible, albeit at the expense of California, Texas, and other large states. Each polygon is a caricature that promotes identification by exaggerating one or more prominent features in its state's outline. Other cues for place identification are the approximate relative positions of the forty-eight conterminous states and the two-character postal codes provided for Alaska, Hawaii, and the District of Columbia. Because identifiable graytone symbols are more important than precise boundaries in the efficient decoding of choropleth maps, distorting size and shape is a useful strategy, especially if the areal units are fa-

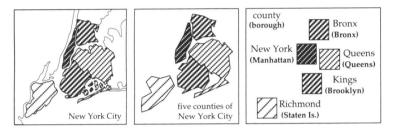

FIGURE 6.13. Three approaches to the design of a New York City detail inset for a county-unit map of New York State.

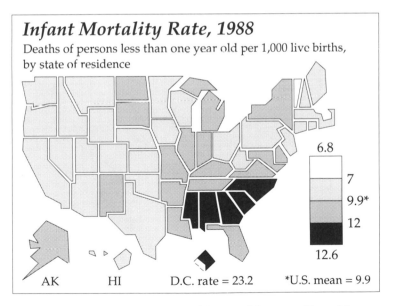

FIGURE 6.14. A visibility base map of the United States provides satisfactory graphic platforms for Rhode Island, Delaware, and other smaller states.

miliar to most viewers and small places on the map are too widely separated for a single detail inset.

Cartograms as Base Maps

Area cartograms extend the purposeful distortion of geography further by making the size of each symbol proportional to the place's population, employment, number of Electoral College votes, or whatever else is being measured. The rationale for the area cartogram is simple: the relative size of the graphic symbol should reflect the relative importance of the areal unit. This concept is particularly appealing to political scientists, who are well aware that some comparatively small states such as Massachusetts and New Jersey wield more political power than large, sparsely settled states such as Montana and Nevada. Maps used in political forecasting often employ a cartogram base to help

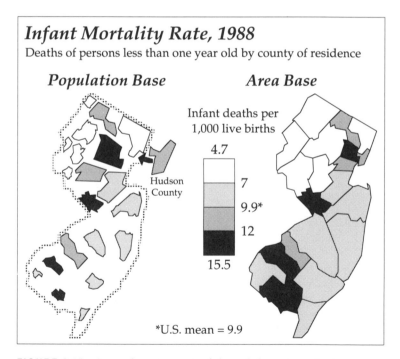

FIGURE 6.15. A complementary pair of choropleth maps uses both an area-cartogram base adjusted for population size (left) and a conventional base (right) to portray the geographic pattern of infant mortality rates.

viewers develop a more accurate appraisal of regional trends.

The left part of Figure 6.15 uses the infant mortality rates for New Jersey counties to illustrate how an area cartogram can help the viewer assess the relative importance of area symbols on a choropleth map. In this example, symbol area is proportional to 1990 population. The cartogram base enhances awareness of tiny Hudson county in the east; in contrast, the symbols of sparsely populated counties in the south and northwest are noticeably smaller than on the conventional base map. Comparison with the conventional choropleth map on the right demonstrates how large, thinly populated areas can have an enormous influence on the viewer's perception of pattern. Although a choroplethic cartogram can be difficult to use without a conventional choropleth map by its side, the cartogram offers additional information useful for interpretation of both individual values and regional patterns.

Making an effective area cartogram is potentially a tedious, iterative process.[5] The example in Figure 6.15 is a *noncontiguous-area cartogram*, in which areal polygons are pulled apart from their neighbors but retain their shapes. Mapping software usually provides an initial solution in which the areal unit with the greatest density remains the same size and all others shrink. The map author can experiment with larger scaling factors and move areas to avoid overlap, and can usually improve the design by providing discernable graytones for a larger share of the map's smaller polygons.[6] Another, more laborious strategy is the *contiguous-area cartogram*, which greatly distorts shape because areal units remain joined along common boundaries. I would recommend the contiguous-area cartogram only for fifty or fewer areal units, only if these units are well known to the map audience, and only if the map author derives masochistic satisfaction from endless tinkering. In most cases, a noncontiguous-area cartogram, or even choropleth shading applied to the interiors of graduated circles, yields an equally effective graphic adjustment with considerably less effort.

Disaggregated Data and Refined Measures

A close look at the national map in Figure 6.14 reveals a smaller range of infant mortality rates among the fifty states (6.8–12.6) than among the twenty-one New Jersey counties (4.7–15.5). The District of Columbia's comparatively high rate (23.2) requires treatment as an outlier, and the national map's single symbol for New Jersey masks considerable variation within the state. At a lower level of aggregation, we may infer quite correctly that some cities and townships in New Jersey have rates much lower than 4.7, while others have rates markedly higher than 23.2. It is a law of geography that smaller and more homogeneous places tend to have a broader range of values than larger, more hetero-geneous divisions of the same region. As a result, the serious scholar must be aware of the averaging and dampening effect of areal aggregation, and sometimes must make maps at different levels and share a few of these with the reader.[7]

Disaggregation of categories in the data might be more impor-tant than disaggregation of areal units. Figure 6.16 demonstrates that the infant mortality rate has very different geographic pat-terns when it is mapped separately for New Jersey's white and nonwhite populations. The two maps and their number-line plot indicate substantially higher rates for the nonwhite population, most of which is African-American. No county has a white rate higher than the statewide average for nonwhites, and only one county has a nonwhite rate below the statewide average for whites. Although the nonwhite rate is roughly correlated with the white rate, racial differences are particularly pronounced in Mercer and Union counties, which have nonwhite rates above the statewide nonwhite mean and white rates below the white mean. Clearly race is an important factor in unraveling and un-derstanding geographic variation in infant mortality.

Race is not the only possible basis for disaggregating catego-ries. A sociologist, for example, would want to examine more detailed dimensions of ethnicity by mapping infant mortality rates for the state's African-American, Asian, Hispanic, and non-Hispanic white populations, among others. Income, education, and other aspects of socioeconomic status would be important as well. An economist examining trends in manufacturing would most certainly want to look at various types of industries to dis-tinguish factors affecting the printing and publishing industry, for instance, from those affecting the chemical industry. But as Figure 6.16 illustrates, disaggregated data might not be available for every place or for categories that are refined as much as the scholar would like.

Other refinements include averaging data for several years and including adjacent places outside the study area. Temporal averaging can dampen minor fluctuations and avoid spuriously high or low rates for small subpopulations. Including adjacent counties in neighboring states is especially useful for maps concerned with employment trends, retail trade, and other phenomena affected by long-distance commuting and shopping trips.

Quantitatively skilled social scientists might attempt a variety of more advanced refinements beyond the scope of this book.

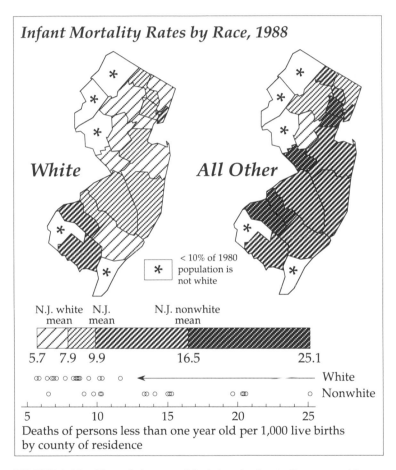

FIGURE 6.16. Choropleth maps of the infant death rate disaggregated by race. My source for data on numbers of births and infant deaths, *Vital Statistics of the United States*, did not provide a racial breakdown if less than 10 percent of the county population was nonwhite.

For instance, residuals from regression might yield a map that adjusts for the effects of urbanization or poverty.[8] A polynomial trend surface or another geographic smoothing technique might separate a mapped pattern into a map of broad regional trends and a map of local anomalies.[9] These and other advanced adjustments or refinements of the data could benefit from complementary pairs or groups of carefully coordinated maps reflecting meaningful categories and graphic logic.

Frame-Rectangle Symbols and Value Estimation

Figure 6.17, the final pair of complementary maps, addresses the estimation of values for individual places. It juxtaposes a choropleth map with a map of frame-rectangle symbols. The choropleth map provides a view of regional trends, and the frame- rect-

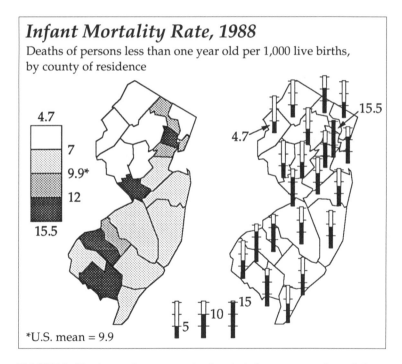

Infant Mortality Rate, 1988

Deaths of persons less than one year old per 1,000 live births, by county of residence

4.7
7
9.9*
12
15.5

4.7

15.5

5   10   15

*U.S. mean = 9.9

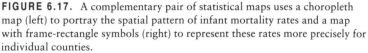

FIGURE 6.17. A complementary pair of statistical maps uses a choropleth map (left) to portray the spatial pattern of infant mortality rates and a map with frame-rectangle symbols (right) to represent these rates more precisely for individual counties.

angle map provides comparatively accurate estimates of infant mortality rates for individual counties.[10] Although an outline map with numbers inserted for each county would be even more precise, the frame-rectangle map supports ready graphic comparison of data values and moderately efficient visual detection of high and low counties. If approximate values of individual rates might be important to the viewer, a frame-rectangle map is a useful supplement to the choropleth map. In this sense the frame-rectangle map can assure the skeptical viewer that the choropleth map does not present a spuriously unrepresentative geographic pattern.

THE ETHICS OF ONE-MAP SOLUTIONS

The examples in this chapter demonstrate that any single map is but one of many cartographic views of a variable or a set of data. Because the statistical map is a rhetorical device as well as an analytic tool, ethics require that a single map not impose a deceptively erroneous or carelessly incomplete cartographic view of the data. Scholars must look carefully at their data, experiment with different representations, weigh both the requirements of the analysis and the likely perceptions of the reader, and consider presenting complementary views with multiple maps. In particular, social scientists mapping quantitative data should consider such complementary pairs as: (1) graduated-circle and choropleth maps presenting a variable as both a count and a rate; (2) a choropleth map presented on both a traditional and an area-cartogram base; (3) a choropleth map supplemented with a more numerically precise frame-rectangle map; and (4) maps comparing more revealing disaggregated categories. Because of inexpensive but effective computer technology, ethically questionable one-map solutions are no longer defensible even as pragmatic expedients.

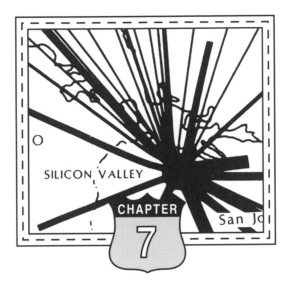

Mapping
Movement,
Change, and
Process

CHANGE, MOVEMENT, AND DYNAMIC GEOGRAPHIC processes in general are difficult to portray on paper—they can be more accurately represented on film or videotape. Animated graphics, however, are not easy to understand and remember, although they may be truthful and intriguing. All but the simplest animated maps can befuddle or alienate a heterogeneous audience, because individual viewers need to control the speed and direction of the display—slowing down, freezing, or backing up and replaying what seems interesting or perplexing in order to relate what they see to prior knowledge and to develop a mental summary of what the animation represents. Although "experiential cartography," supported by interactive electronic graphics systems, can be a captivating training aid, a serendipitous exploratory device, and a powerful design tool, the desired result often is a comparatively straightforward static display that makes the dynamic reality comprehensible. Indeed, my experience with computer-generated animated maps suggests that a rich and lively sequence of cartographic images is truly effective only if summarized by one or two simplified static graphics. Maps in books and journals can convey just a limited sense of dynamism, but the static medium in no way prevents the map author from capturing and communicating the essential facts and relationships of a dynamic process.

Humanists and social scientists who need to map dynamic phenomena should be aware of the variety of cartographic strategies for portraying spatial change or supporting a geographic narrative. This chapter identifies several typical problems in temporal mapping and describes general solutions the map author might easily adapt. Many of the examples demonstrate the need for complementary pairs or triads of maps to portray significant components of change over time and make the information accessible to the reader.

MAPPING FLOWS

Flows—of people, products, or information—often seem to beg for cartographic portrayal. It is difficult to describe with words alone the geography of explorations, invasions, refugee movements, cultural exports, or laundered money, and making a map can be the litmus test of whether an author really has hard facts to back up sweeping generalizations. The flow map has an intriguing elegance; the scholar with relevant data often cannot resist its ability to organize information and capture the reader's attention.

Flow maps can be as simple or complex as flows themselves. A simple straight-line arrow can link an origin and destination, or a curving line with arrowheads to show direction can provide a more exact description of the actual route. Multiple arrow or directed-line symbols can illustrate separate flows or stepwise flows that link an origin, one or more intermediate destinations, and an ultimate destination. Arrows or flow lines can vary in thickness to show differences in magnitude, or can vary in color or pattern to show qualitative differences. Solid and dashed flow lines can differentiate routes known with certainty from routes reconstructed through inference. Flow symbols also can bifurcate to show flow patterns that split or merge. A busy, complicated flow map can demonstrate, for instance, not only the complexity of movement but the meaningful geographic variation in the density of a transport network.

International trade, interregional migration, and other complex flows often require looking only at selected flows or using more than one map. Strategies for simplifying flow maps include showing only extraordinarily strong flows and focusing on a specific origin or destination. Extraordinary flows can be identified by ranking according to volume and then drawing arrows between the origin and the destination for the top 1, 5, or 10 percent of all flows. Demographer Larry Long examined historical shifts in American domestic migration by comparing maps of the five largest interstate flows for various periods.[1] Figure 7.1, adapted from a larger series used by Long, contrasts migrations out of Oklahoma and into California in the late 1930s with more recent streams of Californians seeking amenities farther north along the Pacific Coast in the late 1970s. These maps also show the persistence of movement from New York State to New Jersey and the strong New York-to-Florida migration stream, which had emerged by the 1960s.

Maps focused on one or more representative origins or destinations can provide visually comprehensible and analytically straightforward case studies. As Figure 7.2 illustrates, single-focus maps also allow the use of straight-line symbols, which are both easier to draw in various widths and easier for viewers to compare. In this example, regional planner Robert Cervero used "desired lines" to show how a housing shortage and inflated real-estate prices in the San Francisco Bay area require many

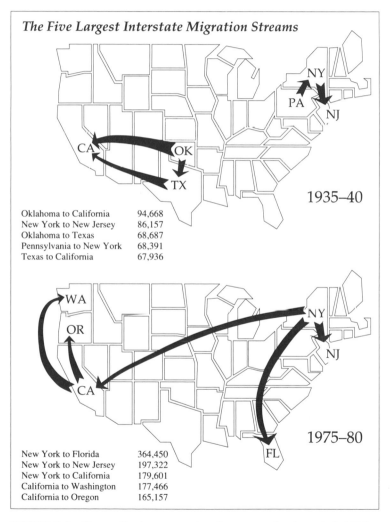

FIGURE 7.1. The top five interstate migration streams for the periods 1935–40 and 1975–80.

workers in Silicon Valley to commute long distances.[2] Cervero used similar maps to compare various destinations and forecast future congestion in northern California.

The data often suggest what maps to use. In a study of interstate flows of medical-school graduates, for instance, my coauthor and I quickly recognized that only two maps were needed: one focusing on California, which emerged as a national mecca for new M.D.s, and the other showing pairs of origins and desti-

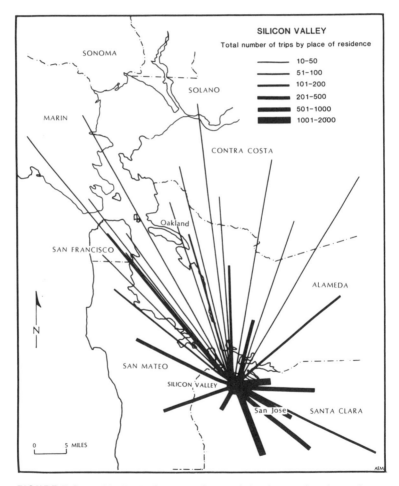

FIGURE 7.2. "1981 Desired Lines and Recorded Volumes of Daily Work Trips from the Entire Bay Area to the Silicon Valley."

nations with significantly large flows.[3] Multiple-map strategies not only yield simplified, more comprehensible graphics but also make it easier to tailor each map to the verbal discussion by identifying important places, features, or routes.

Multiple-map strategies require the author to select specific themes or aspects of a flow phenomenon that merit cartographic portrayal. In a study of migration, for instance, it is often important to distinguish areas of net gain from areas of net loss. If the size of gains or losses is important, graduated point symbols can show absolute size on one map, and choropleth or frame-rectangle symbols can show relative size as rates on a second map. Specialized ratios are also useful. For example, the Bureau of the Census examines commuting behavior with a county-unit map showing the ratio of workers to adult residents; this map identifies "bedroom counties," which have significantly more working residents than workers, and important employment centers, which have significantly more workers than working residents.[4] Studies of commodity flows might require adjusting the data for intervening distance or transport cost, production capacity (supply) at the origin, or consuming capacity (demand) at the destination.[5] But whatever measures are employed, the map author usually must design a specific map either to describe groups of places as major buyers or sellers or to highlight important linkages between pairs of places.

SPATIAL-TEMPORAL SERIES AND MAPS OF CHANGE

Often the scholar can observe a dynamic process only by working with a series of single-date cartographic snapshots based on annual surveys or periodic censuses.[6] Even when more continuous data are available, a set of separate maps evenly spaced in time allows the reader to assess the general direction and pace of change, as well as to examine the geographic pattern of high, medium, and low values or densities for individual dates. Juxtaposition is desirable, because the eye needs to jump freely from one map to the next in order to detect differences. A creative layout based on the smallest effective scale, omission of irrelevant peripheral territory, and careful selection of key dates might allow all maps in the series to be grouped on a single page or on two facing pages. To make this type of display effective, the map author must also standardize scale, format, symbols, and classification so that the viewer only needs to work with a single code. For example, the same medium-gray area symbol should represent the same range of ratio or percentage values on all choropleth maps in a series, so that the viewer can infer correctly that a state or county with a darker symbol for 1950 than for 1940 did indeed register a net increase during the decade.

Maps in a temporal series are especially useful for describing the spread or contraction of a distribution. As an example, Figure 7.3 illustrates how urban policy analyst Phyllis Kaniss used a pair of juxtaposed choropleth maps to demonstrate the contraction of the *Philadelphia Inquirer*'s market area between 1962 and 1988.[7] Kaniss's study used map pairs for the *Inquirer* and for four other metropolitan newspapers to show how television and strong small-city newspapers have eroded the influence of big-city daily newspapers beyond the immediate suburbs. Figure 7.3 also illustrates how juxtaposed maps with a single key and main title conserve space and allow larger type to be used to focus attention on the various periods or dates represented.

Cartographic symbols focusing attention on areas of growth or decline are often useful. Three or more maps representing specific time periods might describe an areal distribution or a network that has continually expanded, such as European settlement in North America, or continually contracted, such as the American railroad network since about 1920. Each map in the series might use a consistent symbolic code to differentiate the cumulative distribution at the beginning or end of the period from the more recent or earlier gain or loss. For six or fewer periods, the map author might conveniently collect all changes onto a single map and use logically coded graytone or patterned symbols to indicate the periods during which the various additions or subtractions occurred; the map then portrays time as an explicit measurement. If change occurred in large, discrete chunks, as it did for the territorial growth of the United States and for large cities that grew in spurts through annexation, exact dates and descriptive labels might replace the key and identify individual additions, as in Figure 7.4, which describes the territorial growth of the conterminous United States. Variations of this map appear frequently in history textbooks and government publications on land management.

For areal units such as states, counties, or census tracts, statistical maps can report change either in absolute quantities or as rates. As indicated in chapter 6, map authors sometimes need a complementary pair of views—a map with graduated point symbols to portray magnitudes and a choropleth map to show relative intensity. When mapping rates of change, a social scientist might appreciate the complementarity of a pair of choropleth maps, one showing gains and the other losses. Figure 7.5 demonstrates how separate maps can emphasize regional patterns: the viewer can see at a glance that during the twentieth century the

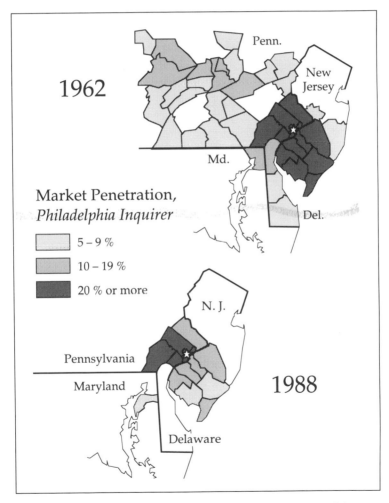

FIGURE 7.3. Juxtaposed pair of maps comparing the market area of the *Philadelphia Inquirer* in 1962 and 1988. Market penetration is measured as the percentage of households within each county receiving the newspaper.

number of daily newspaper firms generally increased in the South and West and declined in the Northeast. Separating gains from losses on separate maps also provides greater detail about relative rates; if presented on a single choropleth map, the information in Figure 7.5 would require a graphically troublesome nine-category graytone scale. If color printing was available, of course, the two maps could conveniently be merged into one, with red area symbols representing gains and blue area symbols representing losses.

Because maps can express change in a variety of ways, not just as amounts and rates, the map author should consider nontraditional measures or themes that might be especially meaningful to the reader. For instance, a rural sociologist or demographer might differentiate areas of recent growth from areas of progressive decline by mapping current population as a percentage of each county's or township's maximum population. Similarly, a historian or historical geographer might approach the demographic history of the American West with a county-unit choropleth map showing the census year of maximum population—and catch the interest of readers unaware that many agricultural counties peaked in population before 1900.[8]

Maps can also summarize mathematical analyses of spatial

FIGURE 7.4. Accessions of territory to the conterminous United States, 1803–1853.

time-series data. A quantitatively oriented researcher might apply a model of temporal change based on exponential or geometric growth, economic theory, or the demographic transition model, and might provide complementary maps of a coefficient describing the pattern of change, with a goodness-of-fit index

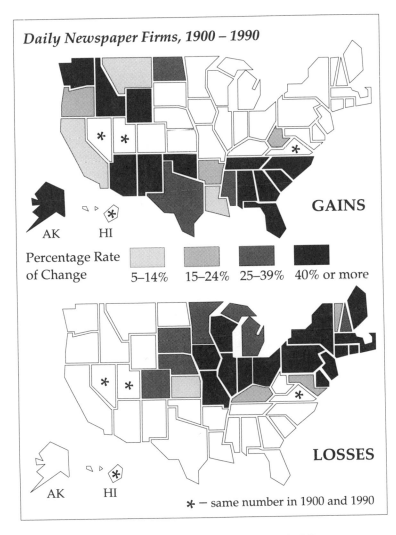

FIGURE 7.5. Separate maps for gains and losses highlight differences between the Northeast and other parts of the United States.

describing how well the model fits the various townships, neighborhoods, tribes, plants, or production units in the study region, for a series of years.[9] Or the researcher might apply several models of change and make a map showing the model that best describes each area or point feature. When comparing change for two successive time periods, the researcher might consider an array of several small two-category maps, each with a straightforward subtitle such as "Accelerating Growth," "Decelerating Growth," or "Increase, then Decrease." Interpretative text in the title or key of a multiperiod or multifactor map can be especially useful in making the results of a numerically complex analysis understandable to readers.

DISTANCE CARTOGRAMS AND RELATIVE SPACE

Scholars examining historical change in transportation rates, interregional migration, or commodity flows can sometimes create dramatically effective maps portraying distances in "airline space," "migration space," or some other *relative space*. The geographer's concept of relative space is based on the straightforward notion that a pair of places with a relatively low transportation cost or comparatively swift transit service is much closer than a pair of places with a higher transport cost or poor transit connections. Indeed, the pair of well-connected places might be much farther apart in physical geographic space, with distance measured in miles or kilometers, than the less well-connected pair. In this sense, a transportation system defines a functional space in which relative distance is measured in money or time. Similarly, flows of tourists, migrants, telephone calls, commodities, or bank transactions also define relative spaces in which pairs of places that interact more are closer than pairs that interact less. Because transport rates, networks, and flow patterns change over time, a before-and-after set of relative-space maps is a useful way of examining the effects of rate deregulation, network improvements, or regional changes in competitiveness or perceived attractiveness.

Juxtaposing distance cartograms comparing two or more time periods is an effective device for demonstrating change in a relative space. The example in Figure 7.6 compares air fares just before deregulation of the U.S. airline industry in 1978 with fares in effect thirteen years later. Each cartogram portrays the least expensive round-trip fares from Syracuse, New York, to ten selected cities. To provide a reliable comparison, both cartograms represent fares advertised for the month of January, and prices on the 1978 cartogram reflect adjustment to 1991 dollars based on Consumer Price Index data, published in the *Statistical Abstract of the U.S.* and the *Monthly Labor Review.* When the cities are arranged according to a common scaling factor, they are noticeably closer to the center on the 1991 map than on the preregulation map. Contraction of the relative space between 1978 and 1991 suggests that deregulation has indeed worked, especially in the competitive, low-demand post-Christmas period, for passengers traveling between Syracuse and Los Angeles on economical but restrictive advance-purchase tickets.

As Figure 7.6 suggests, relative-space maps are easy to both construct and discuss when each map focuses on a single origin or destination.[10] The map author frames the map as a series of concentric circles and labels selected circles to provide a key. Point symbols identified by labels are positioned around the focal point. Because labels must not overlap, relative directions are only approximate. Mentioning the focal origin or destination only in the title avoids confusion and graphic conflict.

When two or more distance cartograms compare different

Round-trip Air Fare from Syracuse in 1991 Dollars

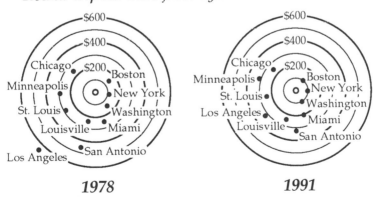

1978   **1991**

FIGURE 7.6. Distance cartogram comparing least-expensive excursion airline fares in January 1978, just before deregulation, with fares in January 1991. Fares for January 1978 have been adjusted to the value of the dollar in January 1991 using the Consumer Price Index.

dates, measurements must be identical, so that changes in distance from the center reflect changes in the relative space. In Figure 7.6, the main title for the overall figure identifies the relative space and individual subtitles identify the date for each component cartogram. It important to standardize monetary values, using the Consumer Price Index, if substantial inflation occurred during the period represented.

Distance cartograms are also useful for comparing different relative spaces, such as air fares and telephone rates.[11] For these comparisons, subtitles might identify the data for each component cartogram, while the main title includes the date. If measurements differ in type of unit (say, travel time in hours compared to travel cost in dollars) or in the range of values (say, $0.56 for a typical one-minute telephone call compared to $150 for a typical airline ticket), a meaningful comparison requires coordinated scaling of the component cartograms. One strategy is to adjust the scales so that the most distant point on each cartogram lies a uniform distance from the center. Other approaches include adjusting the scales to hold constant the average distance of points from the center or to equate distance from the center to a key origin or destination.

FRONTS AND FRONTIERS: MAPPING WAR AND SETTLEMENT

A military occupation of hostile territory usually yields a mappable line of defense. If this line is temporary, military strategists and politicians might optimistically call it a *front*. If opposing forces have negotiated a cease-fire, it's termed a *truce line* or *armistice line*. A wide and somewhat ill-defined military line might be called a *zone*, whereas a longstanding, well-marked, and heavily fortified line is merely a *border*. When settlers as well as troops encroach on territory formerly or still partly inhabited by a weaker, less technologically advanced group, the line becomes a *frontier*.

Historians use line symbols to describe these fronts or boundaries, explain their evolution, and interpret their effects. Because lines of defense or attack often shift frequently during a long battle, historians employ date labels or line symbols that differ in pattern to represent multiple positions of a front. On highly detailed battle maps, line symbols must distinguish not only the opposing forces but also specific divisions, commanders, or regiments. Complex battles with multiple assaults by both sides often yield busy, intertwined groups of battle-line symbols requiring frequent references to a map key. The military historian offering a detailed description of a battle might serve the reader better with separate maps for each important phase of the battle.

Michelin used an effective multiple-map strategy in its 1919 guide to the First Battle of the Marne, which occurred in early September 1914 when French forces thwarted Germany's attempt to capture Paris.[12] Figure 7.7 is one of a series of eight maps in the 18-page historical overview at the beginning of the guide. On each map, solid and dashed lines compare the front at the end of a day of battle with its position at the end of the previous day. Against a background of rivers and town names, the front symbols and the names of commanders provide a straightforward day-by-day cartographic description of who opposed whom where and with what success. Michelin's editors enhanced the guide's visual appeal and eliminated considerable page-turning by positioning each map close to its associated text.

For fighting that moves steadily in one direction, the historian can avoid complexity and confusion by showing the front for only a few key dates. If one side advances consistently, arrows showing major thrusts can be more important than symbols representing

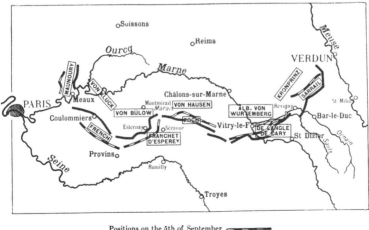

FIGURE 7.7. *Michelin Guide* battle map for September 6th.

fronts. Figure 7.8, a map describing the Allied invasion of North
Africa and Italy during World War II, addresses a much wider area
and period of time than the Michelin map and focuses on strategy
rather than tactics. The map shows only two fronts, the Winter
Line of 1943–44 and the Gothic Line of 1944–45, where Allied
forces dug in and resupplied during the two winters following the
invasion of Italy in 1943. Prominent arrows and informatively
placed but visually recessive date labels describe the northward
progress of the war in the Mediterranean region.

Maps can focus readers' attention on key events. For example,
historian Keith Robbins included only four maps in his short but
wide-ranging essay *The First World War*.[13] Robbins used the map

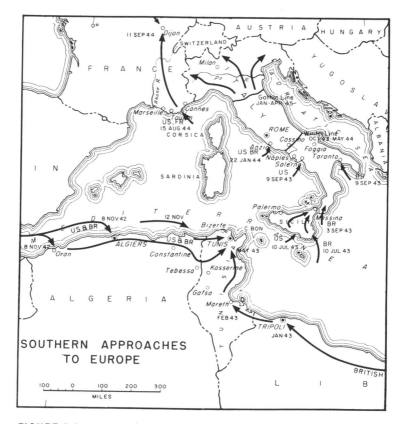

FIGURE 7.8. Portion of the map showing the Allied Army's Southern
approaches to Europe in World War II.

in Figure 7.9 to interpret the Battle of the Marne by showing the thwarted German strategy and the aftermath of the French victory. A thick black line focuses attention on the position of the front just before this major battle. Screened gray arrows show the successful southward advance of German forces, and a dashed gray arrow describes the German plan to encircle Paris from the west. Contrasting white arrows outlined in black show the counterattack of the British and French, and the dotted line represents the position of the front several months later, in early 1915. Robbins demonstrates effectively how a map can summarize a major battle and show relationships among its salient elements.

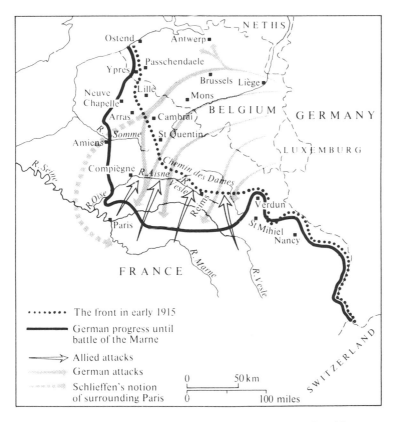

FIGURE 7.9. Map of the Western Front in World War I in Keith Robbins, *The First World War.*

THE MAP AS A NARRATIVE

By organizing information chronologically as well as spatially, maps can support a variety of historical narratives addressing long or short periods of time. A map can organize the critical details of a period of an hour or less, to show, for example, the Kennedy assassination or the key phase of an important battle. Prominent sequence numbers ("1, 2, 3, . . ."), hour-minute time labels ("12:30 p.m."), and directed lines showing routes allow the map to organize events in both space and time. Slower processes, viewed less minutely over longer periods, might include fronts or dates expressed as ranges in days, months, or even years, as in Figure 7.9. For periods as long as a century, a map can summarize, for example, the life of an important diplomat, scientist, or writer. At its simplest, a map might provide a geographic framework for the place names mentioned in a biography. But a more analytical biographical map could relate the subject's travel experience to his or her achievements.

Why biographers and literary analysts rarely use maps puzzles me. Perhaps, as "word people," the idea of providing a geographic summary of an important person's life never occurs to them. Perhaps they merely lack the tools, skills, and self-confidence to attempt a map themselves, or the funds to hire a cartographic illustrator. Or, perhaps by tradition, publishers, critical reviewers, and readers don't expect maps in these genres. I spent a frustrating half day searching our library's collections in vain

Streets opened from 1838 to 1864 and railroad extensions toward the waterfront.

Streets opened from 1878 to 1899 and B&O Railroad tunnels and crossover, built ca. 1893.

FIGURE 7.10. Sections 6 and 8 of an eleven-map cartographic narrative in Sherry H. Olson, *Baltimore: The Building of an American City.*

for an excellent biographical map. Although biographers occasionally provide detailed chronologies, they rarely organize the information spatially. And literary scholars exploring, for example, Dickens's London or Joyce's Dublin are equally reluctant to provide a geographic framework. In contrast, photographs seem de rigueur, in order to assure the reader that the subject was once a child or needed glasses by the age of forty.

Historians of cities, nation-states, railways, and other spatially extensive subjects use maps more readily—sometimes with a grace that both attracts and illuminates. In a study of the economic, social, and political evolution of Baltimore, Maryland, for example, regional scientist Sherry Olson effectively summarized the city's development with a graphic abstract of eleven maps showing rivers, shoreline, roads, and railways.[14] These maps (Figure 7.10 shows two of them) are an integral part of a short, four-page introduction to the biography of a city both nurtured and constrained by its waterfront. Ten of the maps appear at the top and bottom of two facing pages, after an initial page of text, and the introduction concludes with a single map addressing waterfront modifications and the addition of expressways during the most recent period, 1935–79. Standardized symbols and familiar shapes helped Olson's graphic narrative flow smoothly, like good writing, from one map to the next. And descriptive text provides interpretation, helps the reader get the point, and links the maps to each other and to the author's written narrative.

Text is equally important for the single-map graphic abstract. Because standard cartographic symbols are seldom adequate for describing unique and important details, narrative maps tend to have abundant place names, feature names, and blocks of explanatory text. Arranging descriptive phrases spatially, perhaps with the familiar box-and-pointer "balloons" of simple news maps, can usefully complement the linear organization of sentences and paragraphs. Text-rich narrative maps in elite newspapers such as the *New York Times* and the *Washington Post* demonstrate regularly the storytelling power of cartographic illustration. But until advances in computer graphics and electronic publishing enabled journalists to make visually effective maps rapidly and economically, news editors and reporters were as reluctant as many biographers and other scholars to recognize the value of maps in helping readers integrate and comprehend geographically related sets of facts.

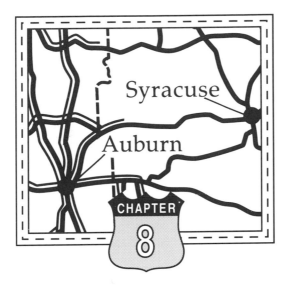

Relational Maps
and Integrative
Cartography

MAPS PLAY A VARIETY OF ROLES, AT DIFFERENT LEV-
els of scholarly discourse. As icons, maps decorate, attract the
eye, and break the monotony of text. As communication devices,
maps help the reader visualize locations, routes, physical landscapes, and abstract geographic distributions. As research tools,
maps help the scholar explore data and search for revealing patterns, relationships, and anomalies. Maps also help the social
scientist test spatial hypotheses and validate spatial models.
These exploratory and confirmatory roles merge with the map's
expository role when the researcher who discovers or demonstrates a meaningful pattern or relationship uses a map to communicate this finding to colleagues and readers.

This chapter addresses the design and use of relational maps
in scholarly writing. Although all maps are relational graphics in
at least a minor sense, some designs are particularly effective in
revealing or demonstrating patterns, trends, and correlations.
The first section discusses the nature of geographic patterns and
examines the use of graphic models for describing, interpreting,
and explaining spatial patterns and trends. The next section explores *cartographic overlay*, a powerful technique for examining
the association among two or more factors. The third section
addresses the representation of bivariate correlation on maps
and statistical diagrams and probes important conceptual differences between geographic correlation and statistical correlation.
And the concluding section explores strategies for coordinating
maps with graphs, words, and pictures, in an integrative cartography in which carefully orchestrated verbal and graphic representations promote understanding by complementing and rein-
forcing each other.

Patterns, Trends, and Spatial Models

Pattern identification can be much more straightforward when
information about places is freed from the linear structure of a
list and displayed on a map. The map allows the viewer to scan

rapidly across a country or continent in search of such potentially meaningful geographic patterns as broad north-to-south or
east-to-west trends, regional concentrations, and urban-rural
differences. In contrast, a one- or two-dimensional table imposes
on the data a comparatively rigid structure that can bias interpretation toward a particular regional or demographic relationship.
Although possibly useful for presenting results, tables are far less
suitable than maps for exploratory analysis.

Mapping the extent and shape of a distribution can help a researcher describe, interpret, or explain a geographic phenomenon. For example, a map revealing linear east–west zones that
decrease in concentration from the equator to the poles might
suggest an important climatic influence. In contrast, a distribution conforming to the shape of a recognized culture hearth invites revision or rejection of a theory touting the dominance of
environmental effects. Because shapes on maps can be serendipitously suggestive, a linear pattern with a distinctive geometry
might demand a careful evaluation of the influences of transport
routes, river valleys, the sea coast, or zones of contact between
feuding tribes. Geographic patterns vary widely in geometry, texture, and meaning, and point patterns that are either more clustered or more uniform than random can also be revealing. Indeed, even seemingly random distributions can be significant,
particularly when a random pattern contradicts a plausible hypothesis. By suggesting visual matches that might confirm hypotheses or suggest explanations, the map exploits a scholar's
accumulated spatial knowledge.

A map that helped the author detect a meaningful pattern can
also help the reader appreciate the pattern's coherence and understand its significance. But because evaluation and interpretation can be distinctly separate tasks, separate maps presented
and discussed sequentially are often more effective than a single
map. Segregating these tasks calls for both a "data map" showing the actual pattern observed and a more abstract map, or
"spatial model," describing the hypothesis or explanation. The
model normally presents a graphically simpler, more easily described pattern than the data map, which reflects a variety of local and often idiosyncratic influences. For an inductive treatment, the data map logically precedes the spatial model, whereas
in deductive discourse the model precedes the map.

A classic example in urban sociology demonstrates how a graphic spatial model can help readers understand a complex phenomenon. Figure 8.1 shows the "concentric-ring" model of urban growth that Ernest Burgess used to describe the changing pattern of land use in Chicago. Burgess, who had looked carefully at Chicago's rapid expansion during the late nineteenth and early twentieth centuries, devised a spatial model to explain and predict the general pattern of expansion for large cities.[1] At the

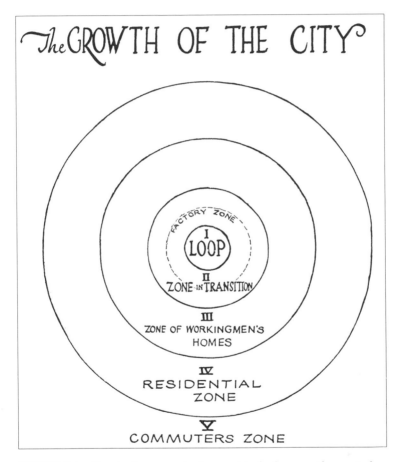

FIGURE 8.1. The classic concentric-ring model of urban growth, a central element in Ernest W. Burgess's essay "The Growth of the City: An Introduction to a Research Project."

center of his model is the commercial district, where railroads, streetcar lines, and important roads converge, as in Chicago's famous "Loop." Major department stores, smart shops, banks, brokerages, and corporate headquarters locate here. Surrounding this commercial center is an expanding zone of wholesale commerce and light industry, labeled the "Zone in Transition."

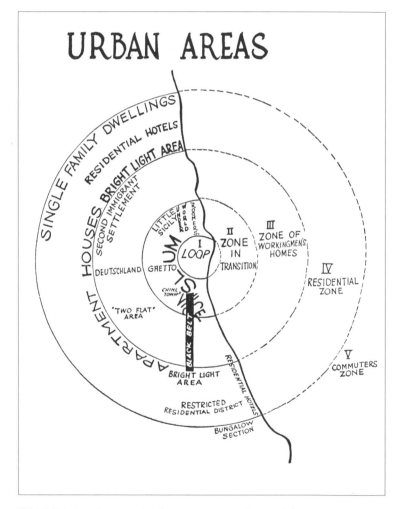

FIGURE 8.2. The generalized map Burgess used to apply his concentric-ring model to Chicago.

The adjacent "Zone of Workingmen's Homes" reflects the need of factory workers to live near their jobs. As the city grows, its zones expand outward, immigrants and other poor families occupy former factory workers' residences in the outer portion of the "Zone in Transition," and factory workers with savings buy single-family homes or rent better apartments in an expanded "Residential Zone." At the outer edge, the city's better-paid managers and office workers commute by rail to the central business district from newer homes on large lots in the "Commuter Zone."

In applying his concentric-ring model to Chicago, Burgess used the illustration in Figure 8.2 to address the difficulty of condensing a large, detailed data map to a page-size illustration in a small book. Chicago's position on the west shore of Lake Michigan is convenient for the map design. The shoreline is an important part of the map's geographic frame of reference: full circles reinforce the concept of concentric circular zones; and labels on the right identify the model's zones, while complementary labels on the left describe specific parts of the city. Burgess used curved labels varying in size to indicate the shape and extent of various districts and neighborhoods and place names such as "Deutschland" and "Little Sicily" to identify important ethnic enclaves.

The concentric-ring model proved to be too simplistic a description of salient trends in metropolitan land use. In 1939 Homer Hoyt, an economist with the Federal Housing Administration, proposed the first major adjustment to the Burgess model.[2] In a highly empirical report richly bolstered by maps and tables, Hoyt recognized the effects on land use of commuter railroads, rivers, and principal thoroughfares. One of the first illustrations in his report is reproduced as Figure 8.3, a map of Chicago's "settled area," defined as all areas with more than one house per acre. The map reflects a radial pattern of railroads crossing largely flat terrain along straight-line routes converging on the city center. This radial pattern led Hoyt to superimpose sectors on Burgess's concentric rings and propose a theory wherein topography and transport routes promote urban growth within well-defined sectors. Figure 8.4, which shows the sector model applied to average monthly rents for six cities, reveals distinct low-value and high-value sectors. As Figure 8.5 demonstrates for Minneapolis and San Francisco, some sectors remained fashionable as new members of a city's elite built or bought homes along established prestige routes. Industry, ethnic groups, and the economically advantaged tended to move radially outward within sectors, as Burgess's labels in Figure 8.2 seem to anticipate. Hoyt used census-tract maps of average rents to demonstrate sector patterns in land values for more than two

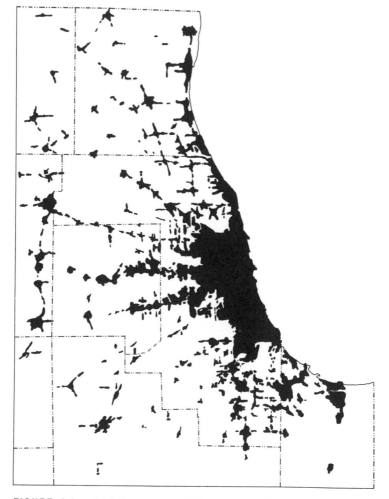

FIGURE 8.3.   Settled-area map, Chicago metropolitan region, 1936, from Homer Hoyt, *The Structure and Growth of Residential Neighborhoods in American Cities.*

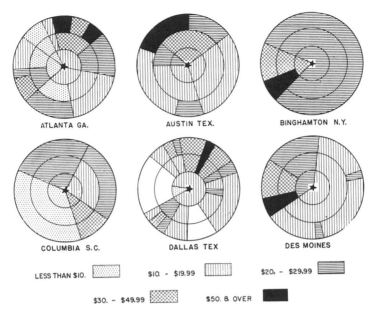

FIGURE 8.4. Theoretical pattern of monthly rents for six cities, 1934.

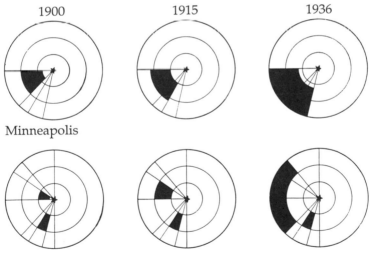

FIGURE 8.5. Shifts in the location of fashionable residential areas in Minneapolis and San Francisco.

dozen cities. Subsequent modifications of the concentric-ring model have recognized the presence of multiple business nuclei within a conurbation, the diminished role of many city centers, and the increased importance of circumferential growth along inner and outer beltways.[3]

Graphic models can be especially useful for interpreting the steps or stages of historical geographic processes. One example is the simple stepwise spatial model that I developed in a short and heretofore obscure paper with the rather tedious title "Railroad Abandonment in Delmarva: The Effect of Orientation on the Probability of Link Severance in a Transport Network."[4] My study area was the Delmarva Peninsula, which includes portions of Delaware, Maryland, and Virginia between the Atlantic Ocean and the Chesapeake Bay. Figure 8.6 shows the form and extent of the area's railroad network for several periods, including 1912, when Delmarva's railroads attained their fullest development. North–south trains linked the area's agricultural centers and coastal resorts to New York and Philadelphia through Wilmington, Delaware, just north of the peninsula, and east–west trains connected at Chesapeake Bay with ferries and steamships serving Baltimore and Washington. As the individual maps for 1869, 1880, and 1912 reflect, the Delmarva railroad network evolved through the superposition of several east–west routes on an earlier, largely north–south network channeling traffic through Wilmington. And as the individual maps for 1944, 1969, and 1991 indicate, even though isolated pieces of these east–west routes survive, the region's present-day network

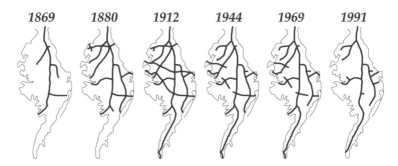

FIGURE 8.6. Railroad lines on the Delmarva Peninsula for selected years, 1869–1991.

owes its basic form to the more stable north–south traffic reflected in the 1880 network.

Figure 8.7 summarizes the Delmarva experience in a three-stage graphic model of railroad abandonment. The first stage represents the fullest development of the railroad network, in which the stronger, more essential lines run from north to south and weaker lines built by speculators run from east to west. Dots vary in size to represent small, medium-size, and large cities. The second stage reflects a substantial decline in east–west traffic and subsequent abandonment of redundant portions of the region's marginal east–west routes. These changes resulted from the Great Depression of the 1930s and increased competition from automobiles, busses, and trucks. Even so, north–south routes remained intact, and the network still served all the original cities. The third stage reflects further pruning of unprofitable branches serving small cities. This simple three-stage model provides a useful generalized explanation of the nature of railroad abandonment in much of the United States. With further modification, a similar graphic model might accommodate such factors as government regulation, weather disasters, state subsidies, tourism, and consolidation of competing north–south routes.

More recently, I used another graphic model to describe and explain the hierarchical geographic pattern of newspaper circulation zones.[5] The model is a two-dimensional, geographic refinement of an even simpler one-dimensional model developed

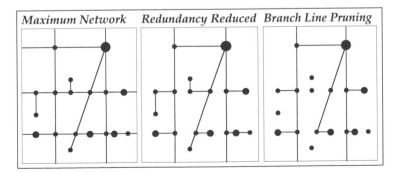

Maximum Network   *Redundancy Reduced*   *Branch Line Pruning*

FIGURE 8.7. Hypothetical stages in the decay of a railroad network in which north–south routes are more stable than east–west routes.

by economist James Rosse.[6] As Figure 8.8 illustrates, at its highest level Rosse's one-dimensional "umbrella" model describes minimal overlap in the home-delivery areas of large-circulation daily newspapers published in central cities. At the next lower level in the hierarchy, several dailies published in satellite cities circulate within the area covered by a single central-city daily, and at a still lower level several nonoverlapping small-city dailies divide the territory of each central-city daily. Circulation zones tend not to overlap among newspapers at a given level, so each paper's principal competition is with a single paper at the next highest level and with several papers at the next lowest level. The territory of a small-city daily thus encompasses several smaller circulation territories, within which weekly newspapers serve towns and villages, at the lowest level in the hierarchy.

To test the Rosse model, I mapped circulation patterns along principal highways connecting small cities in central New York State. Because the daily newspapers use brightly colored delivery tubes bearing their names, home-delivery areas were easy to identify and map. Figure 8.9 shows just the central portion of my "windshield-survey" map, covering an area extending just beyond the circulation territory of the morning and afternoon newspapers published in Syracuse. Thick lines represent delivery routes for the Syracuse dailies, and thin lines represent delivery routes for afternoon newspapers published in Oswego, Auburn, and other smaller cities represented on the map by large black dots.

My data not only verified the umbrella like nesting of the

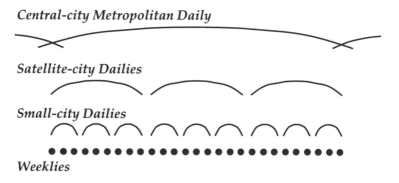

FIGURE 8.8. The Rosse umbrella model of newspaper circulation zones.

Rosse model but revealed other relevant factors, such as a competition-free zone around Syracuse and the effect of county boundaries. For example, a small newspaper that covers county government stories for just a single county tends to circulate throughout its home county but not far into neighboring counties. Yet because readership and news coverage decline with distance, the daily newspaper in a small county is more likely to cover the entire county or reach beyond its borders than a newspaper in a comparatively large county. Moreover, if the city of publication is near the county's border, the home-delivery area

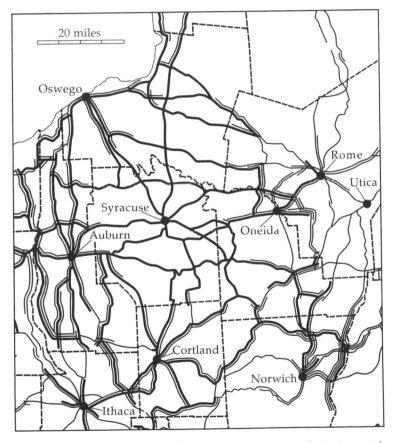

FIGURE 8.9. Pattern of home-delivery daily newspaper circulation in central New York.

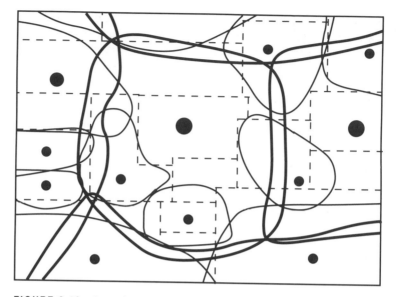

FIGURE 8.10. A graphic summary that extends the Rosse model to two dimensions and accounts for the effects of county boundaries on the shape and extent of circulation zones.

generally will extend farther outward toward the opposite side of the newspaper's own county than it will penetrate the neighboring county. The two-dimensional model in Figure 8.10 describes these effects, complements the field observations mapped in Figure 8.9, amplifies the basic one-dimensional model in Figure 8.8, and demonstrates again how a generalized cartographic model can highlight important components of spatial pattern.

SUPERPOSITION AND ADDITIVE OVERLAYS

Plotting two or more distributions on the same map is a powerful tool in relational cartography. When logic or public policy predict similar geographic distributions for two factors, cartographic overlay can be useful for testing hypotheses or evaluating government programs. Because superimposing two distributions can reveal not only zones of coincidence but also anomalous areas where one factor is present but not the other, overlay analysis might also suggest the need to consider additional factors or to examine some areas in more detail. And if a more complete explanation might lie in not one but a set of several distinct yet closely related factors, overlay is a convenient method, either for considering all relevant factors simultaneously or for assessing the relative significance of each. Indeed, the power of overlay analysis in city and regional planning, environmental science, and public administration accounts for much of the widespread enthusiasm for the sophisticated, computerized geographic information systems (GIS) used widely to manage and integrate cartographic data.[7] Indeed, GIS has become a primary tool for exploratory geographic analysis, largely because it automates map analysis formerly carried out with tracing-paper overlays.

Cartographic overlays vary greatly in graphic complexity, depending upon the number of factors superimposed and their level of detail. Although usually effective when plotted in color on either transparent paper or a computer monitor, cartographic overlays printed in monochrome become visually complex and cluttered as the number of factors increases. Designing a cartographic overlay requires particular care so that graphic interference does not obscure the map's message or pattern.

Among the simplest yet most useful cartographic overlays are maps that relate area features such as residential land to linear features such as commuter railways and freeways. Area symbols must be chosen carefully, though, to maintain contrast when lines cross areas. In particular, dark or coarse area symbols should not overlay line symbols, which are difficult to recognize and follow when superimposed on dark-gray tints or parallel-line area symbols. Of course, visual contrast is easily obtained if only lighter, fine-textured area symbols overlap line symbols. But if the map must portray association between transport routes and high-rate areas, satisfactory contrast might call for surrounding linear features with a narrow buffer of white space. Figure 8.11 shows how demographic historian Barbara Anderson used white-space buffers in an effective cartographic overlay demonstrating the geographic association between the Trans-Siberian Railway and migration to Asiatic Russia in the late nineteenth century. As Anderson notes, high migration rates away from the railway in the lower part of the map reflect the attraction of highly fertile soils in Kazakhstan—an explanatory factor suggested indirectly by her cartographic overlay.[8]

Although superimposing multiple area distributions invites visual clutter, multiple overlays can be an effective way to explain or interpret a distribution influenced by two or more factors. Homer Hoyt, the urban economist who devised the sector theory of urban growth depicted in Figures 8.4 and 8.5, avoided visual clutter by using maps printed on transparent film. In a display titled "The Coincidence of Factors Indicative of Poor Housing, Richmond, Virginia, 1934," Hoyt attempted to explain the spatial pattern of city blocks with housing units averaging less than $15.00 monthly rent.[9] Onto a paper map of low-rent areas, he superimposed four transparent maps printed on film and

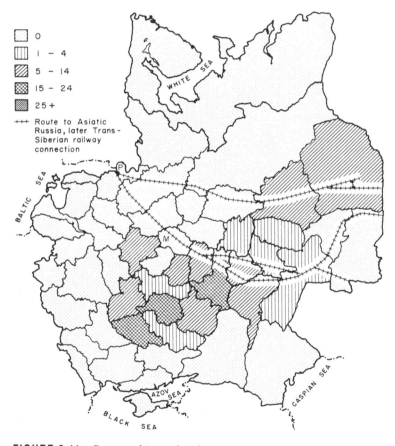

FIGURE 8.11. Cartographic overlay showing the relationship between the Trans-Siberian Railway and migration rates to Asiatic Russia, 1890–94 from Barbara A. Anderson, *Internal Migration during Modernization in Late Nineteenth-Century Russia.*

glued to the paper along their left edges. Separate overlays identify blocks where: (1) at least 25 percent of the buildings needed
major repairs; (2) 75 percent of the buildings were more than
thirty-five years old; and (3) nonwhites constituted more than 50
percent of the population. The fourth overlay shows blocks
meeting Hoyt's thresholds for rent and for all three explanatory
factors. As the portion of Hoyt's composite map in Figure 8.12
illustrates, on each overlay map a different area symbol highlights blocks exceeding the threshold, and separate keys on the
five individual maps combine to form a composite key.

Hoyt's overlays demonstrated the need for a multifactor approach to identifying and explaining areas with the worst hous-

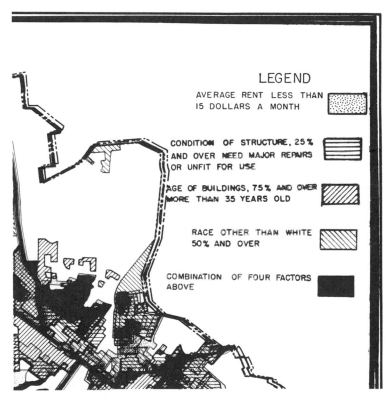

FIGURE 8.12. Facsimile of the upper right-hand portion of Homer Hoyt's
composite cartographic overlay produced by superimposing four transparent
maps on a single paper map.

ing conditions. Although the area covered by his uppermost "combination" layer is much smaller than the entire area of low-rent housing, almost all low-rent blocks exceed the threshold for at least one of the three explanatory factors. And although Hoyt's thresholds were experimental and somewhat arbitrary, his factors were clearly complementary rather than redundant.

Hoyt demonstrated the utility of cartographic superimposition, but his system of transparent overlays was expensive to reproduce and, as Figure 8.12 attests, graphically imprecise. Fortunately for authors who want to use relational cartography in a book or journal article, the overlay technique does not require such elaborate, costly printing. Indeed, Hoyt could have demonstrated his approach more simply and exactly had he relied solely on maps printed on paper. After all, gluing his transparent overlays to the page (and each other) not only anchored them to the book but also only fixed the sequence in which the viewer could add new information and combine factors. Thus, the "condition of structure" map at the bottom of the stack of transparent maps can be superimposed separately on Hoyt's rent map; the "age of buildings" overlay can be superimposed only on a combination of the rent and condition maps, not on the rent map alone. A series of paper maps can present a similar sequence of overlays, and if an author deems separate overlays of single explanatory factors essential, these can be included as well.

The order in which cartographic overlays are added can affect the coherence and flow of the accompanying narrative. Either the logic underlying a hypothesis or the relative explanatory power of individual factors might suggest a specific sequence of explanatory overlays. In the latter instance, the researcher would present first the overlay that best fits the pattern to be explained and order subsequent overlays so that each new factor accounts for a declining portion of the territory not explained by previous factors.

Figure 8.13, a illustrative example of one map in a series of additive overlays, demonstrates how a simple system of area symbols can promote clarity and flow. At the top of the map's graphic hierarchy, the prominent title and dark dot-screen area symbol focus attention on areas where the explanatory factor introduced in the new map overlaps and thereby helps account for the distribution under study. Horizontal parallel-line symbols represent the remainder of the subject pattern, with continuous parallel lines indicating areas where previous factors have accounted for the subject distribution and dashed parallel lines identifying areas yet to be accounted for. Sequences of black-and-white additive overlay maps obviate more elaborate maps printed in color or on transparent material. Although perhaps more numerous than author and editor might like, individual maps in such a series are generally smaller than a single composite map would be and thus could be positioned near their accompanying text.

Superposing maps of geographically separated regions is a useful extension of the overlay concept. For instance, newspapers and news magazines found that overlaying a map of Iraq on a map of the northeastern United States at the same scale helped readers grasp the relative distances involved in the Persian Gulf War of 1991. I have used this technique in preparing for foreign travel and can recommend it highly for tourists wishing to compare distances. In more scholarly endeavors, technological histo-

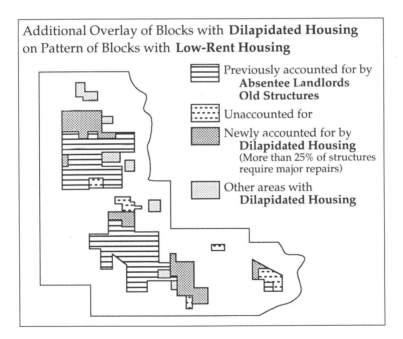

FIGURE 8.13. Typical overlay map introducing a new factor in a hypothetical additive sequence of explanatory cartographic overlays.

rians have found regional superposition useful in examining opposition to standard-time zones proposed by railway officials and daylight-saving-time schemes developed by energy-conscious federal officials.[10] Figure 8.14 illustrates how Ian Bartky and Elizabeth Harrison explained the particular opposition by parents in Cincinnati and Atlanta to the "advanced standard

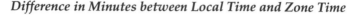

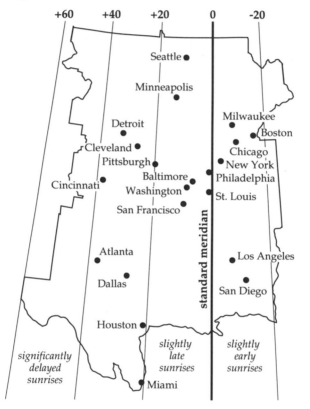

FIGURE 8.14. Superposition of the four time zones for the conterminous United States indicates that most cities with populations greater than 1.5 million are well west of their zone's standard meridian. "Advanced standard time" produced late sunrises in Cincinnati and Atlanta. Composite outer boundary of the four zones indicates that much of the country lies west of the standard meridians for the zones.

time" instituted during the 1973–74 energy crisis. Superimposing the nation's four principal time zones aligned according to their parallels and standard time-zone meridians points out that many American cities with populations of 1.5 million or more were well west of their time zone's center. The sun rises later in these cities than in more eastward parts of the time zones. Switching to daylight time in January to conserve heating fuel delayed sunrise an hour and forced many young children to walk to school in the dark. Social scientists and even art historians might find similar regional overlays useful in accounting for temporal geographic patterns in recreation, voting, television viewing, vocational behavior, or artistic representation.

Representing Geographic Correlation

Bivariate correlation merits separate treatment. Although superimposing two choropleth maps as a form of cartographic overlay is one way of exploring the relationship between a pair of geographic attributes, statisticians customarily employ other devices, most notably the scatterplot and the correlation coefficient. This section compares and contrasts these approaches to representing the association between two factors measured for a common set of places. It demonstrates that geographic correlation and statistical correlation are conceptually different, and that maps, scatterplots, and correlation coefficients are complementary rather than redundant.

A variety of treatments can describe a relationship between two quantitative geographic variables. The simplest approach is the place-wise data table with three columns, the first column containing names or identification numbers for a set of locations, and the second and third columns containing measurements for these places for two attributes. Usually the first column of the table is in alphabetical or numerical order. Although tabular format provides little information about the relationship between the attributes, a conventionally ordered first column promotes the rapid retrieval of measurements for specific places. In contrast, a correlation coefficient describes the relationship between the two attributes with a single number but reveals nothing about specific measurements or their locations. Between these two extremes lie the scatterplot and several kinds of statistical maps.

The scatterplot is an attribute-space representation of a bivariate correlation. Its axes are measurement scales for the attributes in the second and third columns of the data table, and its points represent the enumeration areas or sites listed in the first column. As Figure 8.15 illustrates, the configuration of points in the scatterplot describes graphically the direction, strength, and degree of linearity of the relationship between attributes. As advocates of statistical graphics eagerly point out, a scatterplot is a necessary supplement to the correlation coefficient, because radically different point configurations can have identical correlation coefficients.[11] To illustrate, the two scatterplots on the right side of Figure 8.15 both yield a correlation coefficient of 0.64 but depict markedly different statistical relationships.

Maps have a role in correlation analyses of geographic data, because neither the correlation coefficient nor the standard scatterplot reveals anything about the spatial patterns of the attributes. Yet *geographic correlation* and *statistical correlation*

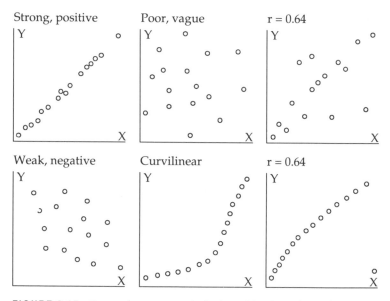

FIGURE 8.15. Scatterplots can reveal whether a bivariate relationship is poorly defined, linear, or curvilinear; whether a recognizable correlation is strong, moderate, or weak; and whether a linear correlation is positive or negative. The two very different scatterplots at the right share the same correlation coefficient, 0.64.

are distinct concepts requiring different analytical techniques.[12] Figure 8.16 demonstrates this point for a hypothetical region with sixteen areal units. A pair of attributes with similar and coherent geographic trends yields a scatterplot identical to that for a pair of attributes with a radically different spatial pattern. The pair of attributes on the left exhibits a strong geographic correlation, whereas the pair on the right does not. Maps that reveal a common spatial trend for two attributes with a strong or moderate statistical correlation can suggest the influence of a single underlying factor. In contrast, dissimilar maps for a pair of attributes with a low or moderate statistical correlation might suggest additional explanatory factors worth considering. Because geographic patterns might prove meaningful in these and other ways, scholars should recognize the inherent complementarity of the map and such nongeographic statistical devices as the scatterplot and correlation coefficient.

Addressing bivariate geographic correlation with a geograph-

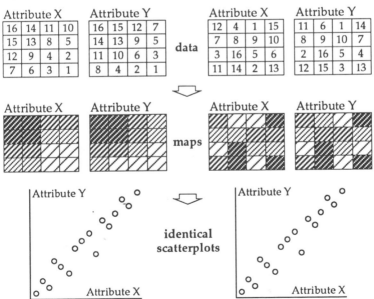

FIGURE 8.16. Pairs of attributes with identical scatterplots can differ markedly in geographic structure.

ic-space graphic raises the question of whether to superpose the two attributes on a single overlay map or to display their patterns separately on a pair of juxtaposed maps. However indecisive it might seem, the best answer is to do both, because separate maps are needed to present what statisticians call the individual "variances" for the pair of attributes, and an overlay map is needed to display their "covariance." Indeed, statistical correlation recognizes all three components in the formula

$$\text{correlation}(X,Y) = \text{covariance}(X,Y) \ / \ [\text{variance}(X) \times \text{variance}(Y)]^{1/2},$$

which divides the covariance of variables X and Y by the square root of the product of the individual variances. Although the mental process of evaluating geographic correlation is far more complex than the standardized arithmetic for the statistical correlation coefficient, a thorough estimate depends on separate—although brief—assessments of each attribute's geographic pattern, as well as on an examination of the attributes' common trends and disparities. As the formula for statistical correlation requires mathematical terms tailored to the concepts of variance and covariance, a thorough cartographic analysis requires multiple maps to address the attributes' geographic trends, both separately and jointly.

Not just any set of maps will do. To promote assessment of the geographic correlation between two attributes, a juxtaposed pair of maps should share a standard design: the same sequence of graytone symbols, the same number of categories, and the same category breaks. After all, if two attributes are geographically similar, their maps should look similar, and if they are geographically different, their maps should look different.

Standardized category breaks are usually based on the *standard deviation*, a widely used statistical measure of a quantitative distribution's dispersion about its mean.[13] Four-category maps are commonly based on category breaks at the mean minus one standard deviation, the mean itself, and the mean plus one standard deviation. Standardized five-category maps require four breaks, usually at ±0.5 and ±1.5 standard deviations away from the mean. Five categories based on the standard deviation conveniently sort places into classes described as "extremely high," "high," "average," "low," and "extremely low"; four categories described as "extremely high," "high," "low," and "extremely low"might produce an abrupt, artificial division at the mean between the "high" and "low" categories.

Unless two geographic distributions are nearly identical, quick visual assessments of geographic correlation based on juxtaposed maps can be misleading. More observations usually fall into the middle category of a five-class map than into any of the other four categories, and if the same places tend to fall into the "average" category on both maps, the mapped patterns can appear quite similar. Moreover, as is true with visual estimates of statistical correlation based on the scatterplot, extremely high or low data values can bias judgments of geographic similarity. Because dark symbols attract the eye more strongly than light symbols, places in the highest (darkest) category have a greater effect on assessments of geographic correlation than places in the lowest (lightest) category.[14] Thus, a pair of five-category maps will tend to look similar if the same places fall into the highest category on both maps.

Figure 8.17 illustrates several additional effects, including a skewed distribution of data values. This pair of maps tests a plausible hypothesis by comparing geographic variances for the female share of elected positions in local government and the female labor-force participation rate. In the upper map, the lowest category is empty, the second category contains nearly as many states as the middle category, and together the second and third categories make up three-quarters of the map's fifty-one areal units. This pattern reflects a distribution mean (20.02 percent) well above the median (19.3 percent), largely as a result of two extraordinarily high values (43.4 and 40.0 percent in the District of Columbia and New Hampshire, respectively). In contrast, the lower map has a more symmetric distribution of data values and looks different from the upper map. This difference is not surprising: the correlation coefficient for these attributes is a modest 0.38.

Variation in the size of areal units calls for cautious judgments of geographic correlation. Although the visibility base map in Figure 8.17 allows the viewer to see Delaware, Hawaii, Rhode Island, and the District of Columbia more clearly than would a standard base map with an equal-area projection, comparatively large states such as Alaska, California, Michigan, New York, and Texas still have more visually influential symbols. Consequently, the highest category does not exert the visual bias mentioned earlier, even though the two areas (New Hampshire and the District of Columbia) with solid black on the lower map are among the four areas with solid black on the upper map. Despite a generally high correlation for New England, all four areas in which females constitute a much-higher-than-average propor-

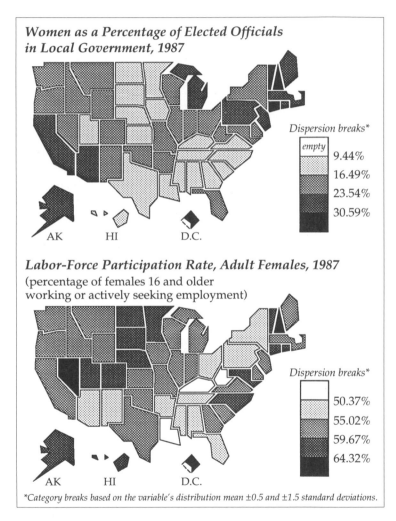

FIGURE 8.17. Juxtaposed maps provide an evaluation of geographic patterns testing the plausible hypothesis that states with proportionately more women in the labor force have proportionately more women serving as elected officials in local government.

tion of elected local officials (Connecticut, the District of Columbia, New Hampshire, and Vermont) are small and graphically weak. If the fifth and highest categories on both maps included Alaska, Montana, New York, and Texas, the visual impression of similarity would be far stronger.

Although hardly similar, the juxtaposed maps in Figure 8.17 reflect several interesting and potentially significant regional patterns. For instance, the upper map reveals greater-than-average proportions of female officeholders in New England and the Southwest, and notably lower proportions in a belt of southern states stretching from Virginia to Texas and in a second block of contiguous states in the upper Midwest. Florida, scarcely typical of the South, has a higher proportion of female officials than its neighbors, whereas New York has a lower rate than any other state in the Northeast. In contrast, the lower map reveals greater diversity within regions, which might explain why states where women tend to work outside the home do not necessarily have higher percentages of women holding public office. The South, for instance, is comparatively diverse, with above-average rates in North Carolina and very low rates in Kentucky and Louisiana. New England is also more diverse on the lower map, and several states in the upper Midwest with relatively few female elected officials have higher-than-average percentages of working females. Moreover, although West Virginia has the smallest percentage of females over age sixteen in the labor force and Nevada the largest, both states fall in the upper map's middle category. By promoting these and similar comparisons, the pair of juxtaposed maps in Figure 8.17 contributes to an understanding of geographic factors that the bivariate hypothesis ignores.

Superimposing two choropleth maps to focus on their attributes' covariation offers a complementary view of geographic correlation. This type of map is called a *cross map* because its key is a cross-classification table, as shown at the bottom of Figure 8.18. The distribution means serve as class breaks for a two-way classification that indicates whether an area is above the mean for both, one, or none of the attributes. The phrase "mean for the states" in the key points out that the distribution means used here are not overall means for the United States but arithmetic means computed by adding up the individual data values for the states and the District of Columbia and then dividing by 51. Symbols and labels in the key are arranged to match the scatterplot in Figure 8.19, where convention dictates that the intersecting vertical and horizontal scales increase toward the top and right, respectively. The use of similar labels, such as "mean for the states," promotes visual integration of the map and graph, as do the thick axes with area symbols similar to those in the map key. With slight prompting by the author, the reader readily recognizes the cross map as the cartographic expression of the 2 x 2 cross-classification formed when distribution means partition the scatterplot into four regions.

As the title of Figure 8.18 appropriately announces, the cross map's theme is the relationship between its two attributes. The graphic logic for the map associates solid area symbols with "correlated," gray symbols with "not correlated," lighter symbols with low data values, and darker symbols with high ones. Thus, the viewer can recognize a strong correlation, if there is

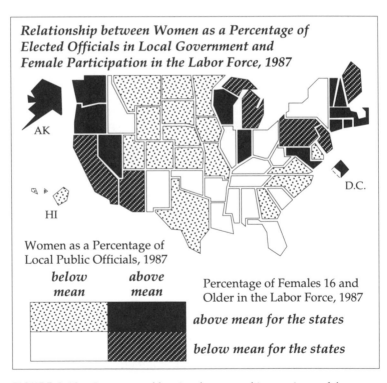

FIGURE 8.18. Cross map addressing the geographic covariance of the attributes portrayed separately in Figure 8.17.

one, as well as distinguish areas with above-average values for one attribute from areas with above-average values for the other. Indeed, if the two attributes had a perfect linear correlation, the trend line in the scatterplot would intersect at the distribution means, and every area on the cross map would be either black or white. For areas with above-average values for only one of the attributes, a gray shade signals a less-than-perfect match with the hypothesis, and the strong visual difference between diagonal lines and dots promotes efficient recognition of the more prominent attribute. Since the female share of elected local officials is

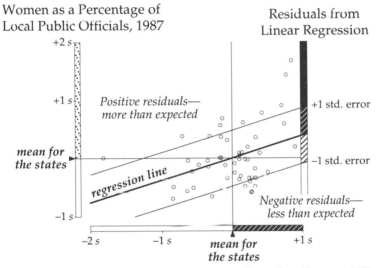

FIGURE 8.19. Scatterplot designed to link the statistical relationship of the attributes to the cross map in Figure 8.18 and the residuals map in Figure 8.20. Vertical and horizontal lines through the distribution means partition the scatterplot into the four regions shown in the key of the cross map. Tick marks on the scatterplot axes indicate standard deviations above or below the distribution mean, and divisions on the residuals axis indicate the standard error units used as category breaks for Figure 8.20. As is typical of regression plots, the vertical axis represents the dependent variable, which the regression attempts to explain, and the horizontal axis represents the independent variable. The three parallel diagonal lines partition the scatterplot into the four categories for the residuals map. The bold diagonal line is the least-squares regression line, and the two thin parallel lines represent residuals at ±1 standard error.

the focal variable for the relationship, this attribute's darker above-average-only symbol allows the viewer to detect readily the generally greater success of female politicians in the Northeast and West.

Although the relationship portrayed is complex, the cross map in Figure 8.18 is easy to understand. Below-average female shares of elective local offices occur in a broad area in the center of the nation stretching from the Canadian border to the Gulf Coast. In roughly half of these states, especially in the Mississippi valley, below-average labor-force participation rates for females support the hypothesis. But in much of the upper Midwest, as well as in such economically developed southern states as Georgia, North Carolina, Texas, and Virginia, below-average female participation in local politics contrasts with above-average female participation in the job market. These states noticeably outnumber states where females are relatively prominent in local government but less likely to be employed outside the home. New England has the largest cluster of states supporting the hypothesis, with above-average rates for both attributes. New York State, often regarded a bastion of liberal and feminist politics, emerges as an intriguing anomaly, perhaps because most of its elective offices are in rural counties outside the New York metropolitan area.

The two-category maps overlaid in Figure 8.18 are not as refined or precise as the pair of juxtaposed five-category maps in Figure 8.17. Of course, it is possible to construct a comparably detailed black-and-white cross map by overlaying a pair of five-category maps, perhaps with vertical parallel-line symbols for one variable and horizontal parallel-line symbols for the other. But the resulting twenty-five-category cross map would be difficult to decode, much less discuss. Color illustration could also be used, to provide intriguing and visually attractive cartographic overlays based on 3 x 3, 4 x 4, and even 5 x 5 cross-classifications, but interpretation would still be tedious, though rewarding.[15] A less expensive and more easily narrated approach might be to identify well-defined or especially significant clusters of points on the scatterplot, highlight their geographic patterns on separate black-and-white maps with carefully phrased titles, and position these maps near the corresponding discussion in the text. In this integrative strategy, outliers could be identified by name on the scatterplot, discussed separately, and mapped only if their locations are noteworthy.

An investigator committed to a more thorough evaluation of a statistical model might want to display the model on the scatterplot and map the "residuals" to show where in the study area the model fits the data and where additional explanatory factors are needed. Since the residual for a data point or enumeration area is computed as the difference

residual value = data value − model value,

an examination of the residuals can be revealing, because places with more of the focal attribute than expected will have positive residuals and those with less than expected will have negative residuals. Whether the scholar needs to show the residuals map to the reader is another matter, of course. But if the relationship is crucial to the author's theme or thesis, and if a careful discussion of the author's analysis would contribute significantly to the reader's understanding and appreciation, integrating a residuals map into the discussion may be well worth the effort required to make the exposition palatable to readers with weak or rusty knowledge of statistics.

Figure 8.19 demonstrates how a scatterplot can link a cross map, a residuals map, and a simple linear regression model. Because it is visually complex at first glance, this diagram requires a careful narrative. The author could begin by examining how the scatterplot complements the cross map, move on to discuss how the regression line generalizes the statistical trend of the data points, and then call attention to data points lying well above or well below the visually prominent trend line. Because each residual is the vertical difference between a data point and the regression line, the magnitude of the residual (that is, its absolute value) measures how well the statistical model fits the attributes for a particular place. Thus, a point above the regression line in Figure 8.19 has a positive residual, indicating that the state's percentage of women officeholders is higher than the level predicted by the regression line and the state's female labor-force participation rate. Conversely, a point below the line has a negative residual, indicating a lower percentage of women officeholders than expected. In addition to establishing the graphic link between the axes of the scatterplot and the key of the cross map,

Figure 8.19 forms a similar graphic link between the residuals scale on the right of the scatterplot and the key of the residuals map in Figure 8.20. Both scales increase upward, from white at the bottom and to solid black at the top, and the thin diagonal lines parallel to the regression line in Figure 8.19 describe the category breaks for the residuals map.

The wordy title of Figure 8.20 reminds readers that the residuals map portrays the geographic pattern of women officeholders, with the effect of women in the labor force removed. Capitalizing the words "NOT ACCOUNTED FOR" indicates pointedly that the map shows residual variation, that is, variation left over or not explained by the model—hence the term "residual." Specific reference to "simple linear regression" is important, because other statistical models can describe the general relationship between these two variables, and the geographic pattern of the residuals will vary from model to model. The residuals map in Figure 8.20 reinforces our examination of the cross map by revealing consistently higher-than-expected percentages of women politi-

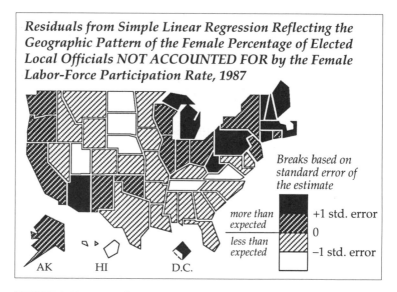

FIGURE 8.20. Map of residuals from the simple linear regression described in Figure 8.19.

cians in local government in the Northeast and Pacific Coast regions and lower-than-expected percentages in the South, the Great Plains, and the Rocky Mountains.

Linear regression is but one approach to detecting and describing bivariate relationships. A social scientist using high-performance interactive statistical graphics software can easily explore a bivariate relationship by fitting a variety of different lines, curved as well as straight. Exploratory data analysis also encourages the researcher to test the effect on curve-fitting of isolated points in the scatterplot with extremely high or low values.[16] Identifying and then omitting one or two outliers might markedly improve the fit between model and scatterplot. Figure 8.21 demonstrates that excluding West Virginia and fitting a fourth-order polynomial describes the scatterplot much better than the regression line in Figure 8.19.[17]

The fourth-order polynomial reveals three regions divided at labor-force participation rates of roughly 55 and 60 percent. In

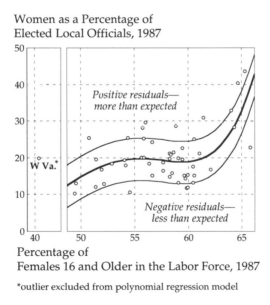

Women as a Percentage of
Elected Local Officials, 1987

*Positive residuals—
more than expected*

W Va.*

*Negative residuals—
less than expected*

Percentage of
Females 16 and Older in the Labor Force, 1987

*outlier excluded from polynomial regression model

FIGURE 8.21.   Scatterplot showing the curvilinear relationship described by fourth-order polynomial regression. A full scale-break isolates West Virginia, an outlier excluded from the polynomial regression.

the left-hand and right-hand regions, the bivariate relationship has a positive, increasing slope, which is especially steep in the right-hand region, whereas in the center the curve is nearly flat, with a slight but insignificant negative slope. Although higher-than-average rates for both attributes on the right and generally lower-than-average rates on the left validate this stairs-and-landing version of the hypothesis, the imperfect fit between model and data suggests that a residuals map might be useful as well.

Figure 8.22 juxtaposes the map of female local officials and the residuals map for the polynomial regression in Figure 8.21. Although the vague similarity of the two maps reflects the limited explanatory power of the female labor-force participation rate, the strong regional pattern of the residuals map is revealing. All six New England states are in the highest category, and the Northeast and Pacific Coast stand out from the remainder of the nation. This pattern suggests the influence of regional differences in culture, politics, and the structure of local government. Town government is stronger in New England than elsewhere, for instance, and some regions offer greater opportunities for citizen participation on elected local school boards and other public bodies. A social scientist exploring this question would surely want to map the per capita number of elective positions in local government.

Cartographic analysis is more enigmatic than the few statistical maps in this section might suggest. Although my illustrations demonstrate how an author can use graphic logic and carefully crafted labels and symbols to link a narrative sequence of maps and graphs, this limited set of examples ignores areal aggregation, ecological correlation, and other relevant methodological issues.[18] Researchers should not assume that state data and county data will yield similar results, for instance, nor should they base inferences about individuals on areally aggregated data. Although data for states and provinces are often revealing, data for counties, townships, city council districts, and census tracts might yield further insights and even contradict findings based on larger, more highly aggregated areal units. Moreover, aggregated data can dilute or mask relationships readily apparent on maps showing individual persons, firms, or establishments. The careful scholar will supplement a broad-brush geographic analysis based on data for large areal units with at least one case study based on individuals or small areas.

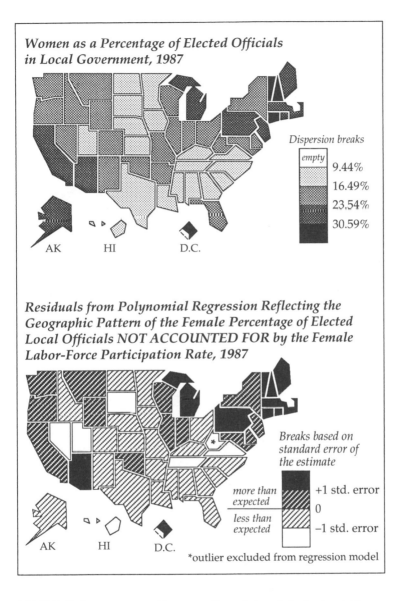

Women as a Percentage of Elected Officials in Local Government, 1987

Dispersion breaks

empty

9.44%
16.49%
23.54%
30.59%

AK   HI   D.C.

Residuals from Polynomial Regression Reflecting the Geographic Pattern of the Female Percentage of Elected Local Officials NOT ACCOUNTED FOR by the Female Labor-Force Participation Rate, 1987

Breaks based on standard error of the estimate

more than expected

+1 std. error

0

less than expected

−1 std. error

AK   HI   D.C.

*outlier excluded from regression model

FIGURE 8.22. Upper map provides an unadjusted view of the focal variable, and lower map presents the geographic pattern of the residuals from the polynomial regression in Figure 8.21. The map keys and area symbols provide visual links to earlier graphics in the sequence, and their dissimilarity reinforces the difference in meaning reflected in the lengthy title of the residuals map.

INTEGRATING MAPS, GRAPHS, WORDS, AND PICTURES

Whereas a lecture audience can simultaneously listen to and watch an oral presentation, a reader receives information through a single, visual channel. Words and sentences structure the author's narrative and, as Marshall McLuhan has argued, their linear format dominates the reader's attention and thought.[19] Scholars in art history, archaeology, and other fields requiring close attention to artifacts have trained themselves to inspect exhibits carefully, but many other readers glance only momentarily at maps and other illustrations. Like many editors and publishers, readers often perceive illustrations as "design elements" inserted to decorate pages, stimulate interest, break up the monotony of column after column of otherwise dull type, and capture the attention of reviewers, casual book buyers, and textbook selection committees. As high-school students, we caught on to this meretricious use of illustration in literary anthologies and social studies texts, and learned to regard it as a bonus. Indeed, if large enough, one picture might even mean a thousand fewer words to read. Consequently, the first goal of integrative cartography is the get the reader to look at the map.

Because the prose already has the reader's attention, the simplest, most direct approach is to say firmly and convincingly, "Look at the map in Figure X." A writer who feels strongly about the research and elects an animated, informal style can forcefully divert the reader. But when an imperative sentence seems too brash or chatty, the most effective substitute is a declarative sentence in the active voice, with the map as its subject. "Figure X shows that . . ." allows the author to weave the map into the narrative by saying why the map is being used and what the reader should be able to see.[20] Putting "Figure X" at the beginning of the sentence in the so-called stress position indicates that the map is an important element in the story and ought not be ignored. A careful choice of words can also help overcome a reluctance to examine illustrations; "demonstrates" is more compelling than "shows," for instance, and few conscientious readers can resist looking when the author says, "Figure X reveals a surprising and significant relationship between . . . ." Even better, particularly for a complex map, might be "Figure X will repay careful study. Among other relationships, the map points out . . . ." In contrast, readers readily ignore illustrations

the writer cites parenthetically. An offhand, footnote-like parenthetical mention such as "... (Fig. X)." at the end of a sentence
signifies that the map is either academic window dressing or dia-
grammatic documentation that the efficient reader can safely ignore. Only slightly better perhaps is metalanguage like "as Figure X shows," tacked on at the beginning or end of a sentence.

Telling the reader to look at a map or announcing the map's
role and significance will probably have little effect if finding the
illustration means moving more than the reader's eyes. Having to
mark the place in the text and turn pages in search of the relevant
map tempts the reader to ignore the map and trust only the author's written interpretation. Hence a second important strategy
in integrative cartography is to place the map near its associated
text, either on the same or a facing page. Although it can be useful to insert instructions such as "Fig. 1 about here" in the manuscript page margins, book designers and production managers
tend to pay more attention to aesthetics than to suggestions from
authors. Visual balance is the prime goal in page layout, so maps
are positioned to make the book look good rather than to make
the reader look at the maps. When map-text juxtaposition is
important, the concerned author should discuss figure placement
with the production manager directly, perhaps in a conference
call including the editor. Early agreement on the size of a book's
type page might encourage the author to develop comparatively
small maps, narrowly focused on relationships discussed in a
paragraph or two, and thus more easily placed near the associated text than a larger map would be, especially one discussed here
and there for several pages. Page-layout software that wraps
lines of text around illustrations helps by encouraging the page
designer to search for visually pleasing solutions that integrate
maps with type.

A third strategy of integrative cartography is to get the text
into the map. As discussed in chapter 4, titles and captions can
underscore the significance of a map by calling attention to the
relationship portrayed or by summarizing the author's interpretation. Balloon symbols similar to those used in newspaper maps
can direct the reader's attention to significant anomalies. And, of
course, the author should talk about the map in the text, perhaps
using relative directions such as "in the lower left" or "just above
the center." If there is little to discuss, the author—and the editor—should ask why a map is used at all. If there is much to dis-

cuss, then by all means discuss it. Even though a well-designed map might be self-explanatory, readers not conditioned to examining illustrations must be told what they can see and where to look.

Names of places and features are an important link between map and text. A locator map, for instance, should show all places mentioned in the associated discussion, as well as a selection of other important places and physical features the informed reader might recognize. When editing or rewriting, the author should carefully compare each map (or a rough compilation) with its associated text. The comparison might suggest other patterns or anomalies on the map worth pointing out in the text and might reveal embarrassing or confusing discrepancies. And when making changes to the map, the author must carefully reexamine the associated text, including later portions of the manuscript that might also refer to the map.

Author and editor should make certain that place names and their spellings match. A map that shows Leningrad or Petrograd, for instance, might confuse some readers if its associated text discusses St. Petersburg. An exception is the facsimile map with archaic labels, for which the author could include either the current or obsolete name in parentheses, as appropriate to the discussion. Discussing differences between an old map and its modern equivalent can be an effective ploy for getting the reader to examine the map.

A fourth strategy is to construct links between successive graphics. Maintaining a comprehensible visual "flow" from one map to the next can be as important as providing a cohesive verbal transition between successive paragraphs. To promote visual flow, integrative cartography can adopt and adapt the expository writing principle of relating new material to old material. After all, what the reader understands after examining Figure X is a useful foundation for the comprehension of Figure X+1. Moreover, old material in the text can support new material on a map, and old material on a map can support new material in the text. Thus, the scholar who consciously and carefully plots an outline that addresses *both* text and graphics is far more likely to produce a well-integrated narrative than one who focuses on words alone and "works the graphics in" as an afterthought. As linguistics expert Joseph Williams demonstrated effectively in *Style: Toward Clarity and Grace*, appreciation of the reader's need for a smooth, logical flow of information is essential for the effective communication of complex relationships.[21]

Conscientious attempts to coordinate graphic sequences and provide coherent symbolic and verbal links might reveal needs that would otherwise be ignored. This chapter has discussed the use of labels, symbols, axes, and keys to promote coherence, the need for multiple maps, and the inherent complementarity of maps and scatterplots. Equally important is the complementarity of maps and other kinds of artwork, including photographs, flow diagrams, process diagrams, and graphic descriptions of conceptual models. Carefully coordinated maps and photographs can be especially useful, for instance, in art history, cultural anthropology, and urban sociology—the map to provide a plan view of interpreted relationships and to show the area covered by the photograph, and the photograph to present a human perspective with important features not represented on the map and thus to remind the reader that the map is a generalization. Improved understanding of places and spatial relationships, and their influence on culture and society, depends upon a free, full, and effective use of all forms of illustration relevant to lucid scholarly discourse.

Appendix A

Drawing Media: Electronic Graphics or
Pen-and-Ink Drafting

Map authors who choose to produce their own maps can use either computer graphics or traditional pen-and-ink methods. This appendix provides a concise overview of both approaches, their advantages and disadvantages, and the equipment needed for three basic map-making operations: compilation, symbolization, and labeling.

Both approaches can be called "manual," because the eye guides the hand in tracing lines and placing symbols and type. In pen-and-ink drafting, the hand holds the pen directly and draws a line by moving the pen across a sheet of paper. In electronic drafting, the hand usually holds a mouse that "draws" by controlling the movement across the screen of a cross-hairs "pen," which leaves in its wake the image of an electronic line. The traditional cartographer typically works directly on the artwork with liquid ink and adhesive-backed symbols and labels. In contrast, the author composing a map on a computer screen obtains a tangible image in a dark powder, or *toner*, fused to the paper by a laser printer, only after plotting all lines, type, and symbols. Although electronic media are more forgiving than traditional media, because the map author can easily make minor changes and print a fresh draft, both media require careful planning and attention to detail.

Had I written this book in the early 1980s, when the success and power of desktop publishing was by no means obvious, this appendix would have focused largely on pen-and-ink tools and techniques. In the late 1980s, though, electronic graphics began to replace traditional media in both cartographic teaching laboratories, where geography majors learn to make maps, and departmental cartographic studios, where staff cartographers prepare maps for faculty. The cost of equipment dropped precipitously, and microcomputer graphics systems became faster and enhanced the user's ability to experiment and revise. Commercial publishers and university presses that adopted electronic systems for page layout and book design could not only accommodate electronic graphics supplied by authors on diskettes, but often preferred them. Although cartography's "electronic transition" has affected most newspaper art departments, commercial map makers, and government mapping agencies, some skilled mapping technicians still use traditional media, which in the right hands enjoy modest advantages in cost and control. And even where electronic media are paramount, manual tools are often used for making minor corrections and adjustments. Persons with excellent hand-eye coordination continue to produce informative, aesthetically attractive maps using technology judged primitive even by the comparatively medieval standards of the 1970s.

I avoid detailed descriptions of either traditional tools or modern software. Appendix C lists a number of excellent references that the interested reader can consult for information about pen-and-ink and photomechanical map production. But because the marketplace for computers and software is changing rapidly, any discussion of specific software packages or hardware products would soon be obsolete. I would rather make readers work a bit harder and find current information than offer misleading, out-of-date facts. Yet I will risk suggesting two valuable and presumably stable sources: the monthly magazine *MacWorld* and the General Periodicals Index and National Newspaper Index databases available on the InfoTrac bibliographic system; they can be searched at a municipal or university library using the keywords "computer graphics" and "maps."

TOOLS FOR COMPILATION

Map compilation usually requires a system for tracing lines and other features and for changing the size and scale of source materials. Although the equipment is radically different, both traditional and electronic technologies support feature tracing and scale change and also provide prepared base maps to which the map author merely adds symbols and labels.

The traditional map maker draws in ink on transparent paper or plastic drafting film, placed atop the source map to be copied. Feature tracing is easier and more accurate, though, when both the source and drafting media rest on a sheet of thick, frosted glass above a bank of fluorescent lights. Back-lighted tracing can be done on a 2-inch thick, portable, and relatively inexpensive *light box*, or on a larger, more costly *light table*, as map makers call a drafting table with a glass top and built-in light source. The hard, planar glass surface also provides excellent support for cutting out the printed point and area symbols sold on self-adhesive sheets.

The electronic equivalent of the light box is the microcomputer system with a scanner and color monitor. The source to be traced is placed face down on a glass sheet, and a sensor travels across the map scanning the intensity of reflected light for tiny cells in an imaginary fine-grained grid with two hundred or more rows and columns to the inch. The scanned image is saved as a picture file, which can be read into a drawing program, placed in a background or inactive layer below the active layer the map author is composing, and converted from black to magenta or cyan. A color monitor provides contrast between the black lines, area symbols, and type in the active layer and the colored images of printed lines and other dark symbols on the source map. For map compilation, a color monitor is superior to a black-and-white (monochrome) monitor, because a map author unable to distinguish lines and type on the new drawing from those on the source image cannot readily determine which features have yet to be traced.

Map authors using electronic graphics have a few less expensive options. Least costly, but generally unsatisfactory, is a sheet of transparent film taped onto the front of the monitor. Although photocopy transparencies are inexpensive, screen tracing is imprecise because the inner, drawing surface of a cathode-ray-tube screen is set back several millimeters from its curved outer surface. A *graphics tablet* with a *stylus*, which temporarily replaces the mouse, is more suitable. The user draws on the screen by following lines across a source drawing placed on the tablet. Because only features transferred from the source appear on the screen, a color monitor is not required. Although a scanner is more flexible and efficient, a patient map author can obtain equivalent results with a graphics tablet.

Traditional devices for changing scale have been either expensive or cheap. University cartographic laboratories have usually had a reflecting projector to allow the map author to copy a source map in pencil or ink at a different scale. In the 1960s and 1970s, as I recall, these machines cost a bit more than the annual stipend for a graduate assistant. Copy cameras, used primarily for making decent-looking final copies of drawings at a somewhat reduced scale, can also be used to enlarge or reduce source materials used in map compilation. Typically, a reflecting projector can reduce or enlarge a 10-cm line to any length between 2.5 cm and 40 cm, whereas a copy camera at its extreme settings might shorten or lengthen the same line to between 3.3 cm and 30 cm. Making the source image any smaller or larger would require two or more successive tracings or photographs.

Less capital-intensive, albeit more tedious and less exact, is the grid-transfer method, in which grids with equivalent numbers of evenly spaced rows and columns are drawn on both the source and the copy. The map maker transfers points at which features intersect grid lines on the source map to their corresponding locations on the enlarged or reduced grid of the copy and then completes the feature by eye, using these transferred points as guides. If a light table or light box is available, a sheet of graph paper or an appropriately scaled grid can serve as an underlay, and grid lines need not be drawn directly on the artwork.

An office copier that enlarges and reduces provides a fast and inexpensive method of changing scale. But xerographic copiers often have a limited range of scaling, perhaps enlarging only to 155 percent and reducing only to 65 percent, so that two or more carefully calculated steps might be needed. Because copiers sometimes stretch or compress the reproduced image in one direction, the map maker should check for distortion by photocopying a sheet of graph paper and comparing horizontal and vertical lengths.

An electronic scanner linked to a computer graphics system can enlarge and reduce across a wider range of scales than an office copier. Because graphics software allows independent horizontal and vertical scaling, the conscientious map author could even remove linear distortion from a xerographic copy made at a map library or archives.

Map authors employing traditional media sometimes add type and symbols to base maps or outline maps printed with a minimal geographic framework. This strategy is especially useful for making choropleth maps by carefully cutting out portions of self-adhesive area symbols and pasting them on the base map. Particularly helpful are base maps with place names and other reference features printed in nonphoto-blue ink, which will not copy when the artwork is photographed. Sometimes useful are simple outline maps containing only a small number of features or labels; the unwanted ones can be easily masked with correction fluid (commonly called "white-out").

The electronic equivalent of this just-add-symbols approach to map making is cartographic clip art, which the map author can easily modify by adding symbols, deleting features, and changing the style and size of labels. Even more important is mapping software that supports experimentation with the design of statistical maps or promotes nearly effortless construction of projection grids tailored to specific areas or distributions. Electronic cartographic media occasionally allow cartographically ignorant users to make misleading but aesthetically pleasing maps that equally naive readers find acceptable. But in the hands of a scholar or artist who understands graphic communication and map distortion, electronic tools can not only save hours of tedious labor but also encourage customized solutions not practicable with traditional media.

TOOLS FOR SYMBOLIZATION

Persons who think of traditional cartography largely in terms of maps drawn with pen and ink are astonished by the variety of symbolization tools and techniques that have been developed for nonelectronic cartographic drafting. As several of the references listed in appendix C describe, lines may be "taped down" onto a drawing with thin, flexible self-adhesive tape or "scribed" with the sharp point or narrow blade of a *graver* into the orange-colored opaque coating of a sheet of plastic film. For a map prepared as "negative artwork," the orange-coated film acts like a photographic negative, and the thin, translucent corridor of uniform width opened by the graver ultimately yields a crisp, uniform line. For a map prepared more conventionally as "positive artwork," the cartographer might burnish onto the paper an appropriately shaped piece of an area symbol cut from a printed self-adhesive sheet. Also useful are graphic arts symbols published on a carbon-coated sheet of *dry-transfer* symbols; the cartographer carefully places the dry-transfer sheet face down on the paper and rubs the back of the sheet with a pencil to transfer the appropriate point symbol, arrowhead, straight line, or alphabetic character onto the correct part of the drawing.

Quality of the artwork depends in part, of course, on the quality of the drawing surface. Although commercial drafting paper works well for pen-and-ink map making, better results are obtained with translucent acetate drafting film, frosted on one side to hold ink and sustain careful scraping with an X-acto knife. Specially coated paper produces a cleaner, sharper image in a laser printer than the standard bond paper customarily used for correspondence and manuscripts (especially important if the map employs small labels and fine-grained graytone area symbols), and photosensitive paper exposed on a high-resolution *imagesetter* yields a superior print.

Despite different drawing surfaces and methods of application, traditional and electronic media are surprisingly similar in the types of symbols they produce. Both media provide lines in a variety of widths and area symbols in a variety of graytones and patterns. Although a poorly equipped studio or workstation might have few of the available options, manufacturers of graphic arts supplies offer a broad range of choices, as do the developers of the better-quality graphics software. And both media also support techniques for adding fine details or correcting minor errors. For example, the pen-and-ink map maker uses a small knife to scrape away unwanted ink and a thin, sharp-pointed "crowquill" pen to add ink, whereas the electronic artist achieves similar results by enlarging the part of the drawing requiring attention and superimposing appropriately shaped symbols, either in black to add detail or in opaque white to hide blemishes.

Although the traditional media support most of what a map author can accomplish electronically, drawing software promotes a variety of practices a map maker or artist would rarely attempt with pen and ink.[1] Two of these are layering and pen patterns for lines, similar to fill patterns for area symbols. Layering allows symbols to be added early or later each time the image is plotted or redrawn on the screen. Because a laser printer plots a drawing on an electrostatic drum before adding toner and transferring the image to paper, the order in which symbols are added is important. A foreground symbol added more recently to the drum can overwrite or obscure all or part of an earlier symbol farther "toward the rear" of the drawing. Thus, a medium-gray-tone line crossing a solid black polygon or area symbol can either be in front of or behind it. Positioning commands allow the map author to restructure a drawing by moving specific symbols or groups of symbols directly to the front, directly to the back, or farther forward or back in discrete steps.

The ability to select and move individual points of a linear feature or boundary is particularly important. Also useful is the ability to move, scale, or duplicate collections of symbols consolidated into a single "object" with the "group" function. As an example, a map author who has drawn a town street grid could generate a double-line road symbol by (1) combining all the streets into a single object, selecting a relatively wide black "pen" for the group, and using the "lock" command to fix its position; (2) creating an identical object with the "duplicate" command, and then choosing a thinner opaque-white pen for this duplicate object and moving it to the front of the drawing; and (3) using the "alignment" command to center this thinner opaque-white duplicate street grid on top of the solid black street grid. This sequence of commands yields a street symbol that appears as two parallel, evenly spaced black lines, as in Figure 3.11.

TOOLS FOR LABELING

Creating decent-looking type is far more complex for the traditional mapmaker than for the map author using electronic graphics. In the pen-and-ink era, a few highly talented map makers saved time and produced pleasing results by painstakingly drawing labels in a distinctive calligraphic script. Less-skilled neophytes who attempted to mimic their efforts scribbled crude labels that announced both lack of skill and lack of concern. More exact, but still prone to tragic accident, was the lettering template used to add labels in ink letter-by-letter directly onto the map. And more forgiving were dry-transfer letters printed on a sheet, to be placed face-down on the drawing and transferred to the paper by burnishing; a letter that didn't look right could be scraped off without disturbing surrounding symbols. Pen-and-ink map makers obtained their best results by cutting labels out of a sheet of type ordered from a print shop, applying a thin coating of wax or rubber cement, and sticking these crisp labels carefully onto the drawing.

Using type ordered from a printer took time and planning, discouraged creative changes, and was prone to occasional typographic errors. Thus, map makers who could afford them purchased machines that provided immediate access and all or some of the aesthetic quality of print-shop type. These machines ranged in price, sophistication, and quality from special typewriters with large letters to "headline-makers" for composing crisp labels letter-by-letter on 35mm photographic paper. Many cartographic workshops acquired their first microcomputers and laser printers as a quick, flexible, and comparatively inexpensive alternative to ordering type from a printer. The map author using conventional media, unless talented in calligraphy, is best advised to compose labels on a microcomputer and print them on "wax-holdout" paper, which accepts wax or rubber cement without marring the type. Dry-transfer labeling, which requires considerable care in aligning and spacing individual letters, is a distant second choice.

Map authors using electronic media have a wide choice of typefaces, styles, and sizes. It is easy to make horizontal labels perfectly horizontal—a once-tedious task that confounded many geography students forced to learn the rudiments of pen-and-ink cartography. Some graphics programs allow curved type so, that labels can reflect the shape of rivers, irregular roads, or curved parallels of latitude. Particularly important when named features are crowded is the ability to experiment directly on the screen with letter size, abbreviations, and nonadjacent positions requiring thin leader lines.

Software Compatibility and Planning

Electronic map making with personal computers has improved markedly since the mid-1980s, but the increased sophistication has its price. As long as microcomputers, operating systems, local area networks, laser printers, applications software, and geographic databases continue to evolve, electronic graphics will remain a minefield of unanticipated incompatibility, unexplained crashes, malfunctioning procedures, and other sources of frustration and delay. New software does not always mesh well with old operating systems, for instance, and haste does indeed make waste, as software developers rush products onto the market with minimal testing. One ought not be considered a Luddite for recognizing that "better" too often is the enemy of "good."

Two principles can help the wary map author avoid frustration. First, don't assume compatibility. When purchasing software or system components, read reviews, describe your system to a technically competent salesperson, and insist upon return privileges. When sending electronic copies of maps and other graphics to a publisher, make certain that the publisher's page-layout software is compatible with your drawing software. If in doubt, submit a sample illustration for testing.

Second, recognize that software can be a limitation as well as an asset. In the mid-1980s, for example, inflexible software encouraged the use of garish, inappropriately coarse area symbols, which became graphic clichés when newspaper artists and other map makers first adopted Macintosh computers and MacDraw. Although software has improved, the available range of mapping techniques and drawing functions can still constrain the form and functionality of cartographic symbols. Never, for instance, use a choropleth map to show variation in magnitude because the only mapping program available is a choropleth mapping application; either borrow appropriate software or use a general-purpose drawing program to make appropriately scaled graduated point symbols. Be aware that some maps will require multiple software. For example, a statistical-mapping application would be needed to generate and scale graduated circles, and a general-purpose drawing application would be needed to reposition some circles, add labels, and make these symbols conform to a standard design. In cartography, as in architecture, let form follow function and shun needless variety.

Appendix B

Working with a Cartographic Illustrator

Either a need for efficiency or a map author's limited skill in working with graphic-arts media might dictate the assistance of a staff cartographer or free-lance cartographic illustrator. Many universities have set up cartographic laboratories or studios to draw maps for geography faculty, and some studios will accept work from outside the department and even from outside the university.[1] In geography departments that have made the transition to computer graphics, the staff cartographer now serves many clients as a specialized consultant or skilled technician who provides advice or instruction in the use of the laboratory's own facilities as well as in graphic-arts, database-retrieval, and computer-graphics services elsewhere on campus. But some clients will still require the more traditional service of a staff cartographer who codesigns and draws their maps. And the free-lance cartographic illustrators found in many college towns and most large cities will still be needed to translate general ideas or careful compilations into finished, publishable maps.

This short appendix examines the working relationships that might exist between a map author and a staff cartographer or illustrator. It also discusses the need for effective communication between author and illustrator.

WORKING RELATIONSHIPS

Map authors work with cartographic illustrators in a variety of ways. At the least intellectually involved level, the author might relegate most design decisions to the illustrator, saying, in effect:

"Here is my text. This is what I want. Assemble the necessary information, make a compilation, and draw me a map. I'll look at it, and possibly suggest some changes." In this instance the illustrator is clearly a coauthor of the map—a "ghost cartographer" if suitable credit is not given. This author-illustrator model is common in journalism, where the cartographic artist plans an assignment with a reporter or editor and is often mentioned by name in a credit line. Like the photojournalist, the editorial artist is not only an experienced technician but also a partner journalist. The artist understands cartographic sources and contributes an informative graphic explanation.

A very different working relationship exists between the scholar who compiles and designs the map in pencil and the illustrator who renders the finished artwork. This might be called the Gary Trudeau model, after the well-known creator of the "Doonesbury" comic strip. Trudeau generates his own narrative, composes text for the balloons, and makes rough sketches of the panels, which he then faxes to an illustrator hundreds of miles away. Although this unnamed illustrator draws Mike, Joanie Caucus, and the other Doonesbury characters, Trudeau is clearly the strip's author. Similarly, the scholar who prepares a map's compilation, layout, and text and specifies its scale, projection, and symbolization is clearly the map author, even though he or she might never touch a pen or mouse. In addition to maintaining aesthetic quality, promoting graphic continuity, and freeing the author for other pursuits, a good cartographer also serves as a "design filter" and "graphics editor," whose contributions vary inversely with the aesthetic skill and cartographic knowledge of the map author. The cartographer is like an experienced copy editor or manuscript typist, in that his or her advice and suggestions can contribute significantly to the clarity and success of the author's creative work.

Electronic graphics has led to new kinds of author-illustrator relationships. Because of powerful software and inexpensive desktop publishing systems, scholars are more willing to assume the lead role in producing finished cartographic art, as well as in compiling information and designing symbols and layout. Although doing it right the first time still has enormous advantages, refining a drawing through minor, incremental modifications no longer carries heavy penalties in time, cost, and frustration. Nonetheless, the map author with limited experience in using the technology and limited access to a scanner, an imagesetter, or professional-quality illustration software needs advice in determining what to buy, where to go, and how much of the map to make independently. Consequently, a staff cartographer or an experienced user of electronic graphics can play a variety of useful roles, from consultant to coach to "finisher" to contractor. As a consultant, for instance, the experienced map maker might recommend software, provide advice about campus facilities or extramural service bureaus, or suggest how-to books more lucid than vendors' manuals. As a coach, the staff cartographer can help the scholar overcome the frustration and uncertainty of working with unfamiliar equipment and new software tools. Coaching duties may include demonstrating how to use drawing software or a scanner, offering helpful hints not found in the manual, and suggesting more efficient techniques. As a finisher, the skilled illustrator might use a professional-grade illustration package to enhance aesthetically a map created by the author with less sophisticated drawing or mapping software. Aesthetic adjustments in graytone area symbols and line widths may be made before printing a more refined imagesetter version of a map developed using laser-printer proofs.

The scholar lacking time or self-confidence to create all or most of the artwork, and unable to afford the university cartographic laboratory's fees, can usually hire a student in geography, architecture, industrial design, or landscape architecture. (Only an independently wealthy author, or one backed by a major grant or a generous textbook contract, should even think of going to a large commercial cartographic firm such as Hammond or Rand McNally.) Before making either a formal or informal contract with a free lance, ask to see a portfolio and request references from satisfied clients, faculty cartography teachers, or a staff cartographer. After all, an illustrator who has rendered many maps may not necessarily understand cartographic principles. In all cases, verify both the free lance's competence in cartographic illustration and his or her willingness to give you, the client, what you want and need.

This is a useful place to remind the author about the different types of publishers and their normal practices in supporting the preparation of maps and other illustrations. A textbook publisher should be willing—and especially eager if it will help sales—to hire a cartographic illustrator to make the maps, whereas a university press would expect the author to provide final artwork. But if the expense is large and the project particularly significant, an editor may suggest one or more foundations to which the author can apply for help. Commercial publishers of nonfiction trade books and scientific-technical reference books usually write into the contract a fixed expense for art preparation, with this cost sometimes treated as an advance against the author's royalties. Because of a strong sense of how well a book will sell, scientific-technical publishers can set a sales price and production budget that make a loss unlikely. Humanists and social scientists, who deal largely with university presses, can benefit from learning as much as possible about the use, design, and execution of maps in order to minimize payments to an illustrator or cartographic laboratory.

What the Illustrator Will Need to Know

A free lance hired to make maps needs information about both the region to be mapped and the book or journal in which the maps will appear. For maps in a book, the author should provide a copy of the publisher's instructions for maps and other illustrations, as well as any specifications developed for the book's design—in particular, paper quality and size of the type page. Paper quality is an important constraint, because fine-grained graytone area symbols and small labels with fine serifs reproduce poorly on coarse paper. Size is important, because the designer or production editor might object to an oversize map intended to extend into the margin or across the gutter as a two-page, "double-truck" illustration. Tipped-in, fold-out illustrations are expensive to print and bind, and turning a map sideways on the page can make layout difficult if a single page cannot accommodate both the rotated map and its rotated caption. An experienced illustrator must understand the publisher's expectations in order to identify potential problems that might require negotiations with the publisher or modifications by the author. For the same reason, the author of a journal article should provide not only a copy of the editor's "guidelines for authors," but also an issue of the journal to illustrate page size and paper quality. If the likely outlets for an article vary widely in page size, all but the most confident author should submit compilations or rough sketches and delay preparation of the finished artwork until the manuscript is accepted, particularly if traditional pen-and-ink media will be used.

The illustrator also requires an appropriate base map of the mapped area. Unless already embodied in the compilation drawing, an up-to-date or historically accurate base map is essential. A single base map covering the entire area on a single sheet with an appropriate projection is preferable to multiple maps covering different parts of the region. If several suitable base maps with different scales or projections are available, bring all of them, or at least a representative sample, for discussion. An ideal base map is one that is in the public domain and is larger, but not too much larger, in scale than the finished artwork. A good rule of thumb for conventional media is to draw a slightly oversize map at a scale 25–50 percent larger than the final artwork and then to reduce this drawing photographically to the required size. Although photoreduction mitigates the visual impact of minor irregularities, too much photoreduction makes it difficult to generalize properly and plan for changes in visual contrast.

A compilation drawing should be detailed and unambiguous, with the region to be shown on the finished map clearly delineated. Colored pencils are useful for distinguishing among various kinds of linear features, as well as for differentiating symbols and labels for the map from instructions to the illustrator. Titles, place names, and other labels should be provided twice: once on the compilation to show their locations, and again on a "type list" to specify the size and style of type required. All labels must be accounted for on both the compilation drawing and the type list. Spelling and wording should be checked carefully; spelling must be not only correct but also consistent with the text. The map author should keep a duplicate copy of the compilation drawing, type list, and list of symbol specifications, and any changes made while discussing the map with the illustrator should be carefully noted on both copies. The map author who cannot communicate a clear sense of the map's objectives, content, and design should be prepared to provide further instructions at a second meeting.

Illustrators are not perfect, and neither are map authors. Working with a cartographic illustrator is an iterative process that requires occasional queries and careful inspection of proofs. Discussion of a project and close scrutiny of proofs often yield new insights so that as the final version of a map evolves, its design and content often change, at least in a minor way. Although normal and usually beneficial, these changes can also be costly. For this reason it is better to arrange to pay the illustrator a flat hourly rate than a fixed price with additional hourly-rate payments for the author's alterations.[2]

Appendix C

Selected Readings

The books listed here may whet the scholar's appetite for further information. Cartographic texts emphasizing traditional media are comparatively abundant; most of the following books present exemplars or discuss design principles useful to map authors working with electronic media.

American Congress on Surveying and Mapping, Committee on Map Projections. *Choosing a World Map: Attributes, Distortions, Classes, Aspects.* Falls Church, Va.: American Congress on Surveying and Mapping, 1988.
——. *Matching the Map Projection to the Need.* Rockville, Md.: American Congress on Surveying and Mapping, 1991.
——. *Which Map Is Best? Projections for World Maps.* Falls Church, Va.: American Congress on Surveying and Mapping, 1986. These three well-illustrated booklets reflect the collaborative wisdom of many experts and offer useful guidance in selecting a projection for a small-scale map of the world, a continent, a region, or a country.
Brunet, Roger. *La Carte: Mode d'emploi.* Montpellier: Fayard/Reclus, 1987. This cartographic catalog covers over eighty different types of maps and themes. Although not all the examples are well designed or carefully reproduced, this rich collection of facsimile illustrations can be a useful cartographic stimulus for environmentalists and social scientists.
Campbell, John. *Introductory Cartography.* 2d ed. Dubuque, Iowa: William C. Brown, 1991. A well-illustrated basic textbook on the principles and techniques of modern map making. Especially useful as a general reference and for its treatment of traditional cartographic drafting.

Clark, Suzanne M., Mary Lynette Larsgaard, and Cynthia M. Teague. *Cartographic Citations: A Style Guide.* Seattle, Wash.: Map and Geography Round Table, American Library Association, 1992. This first citation guide for maps conforms to the general style guidelines of *The Chicago Manual of Style.* Available from Kathryn Womble, MAGERT Distribution Manager, University of Washington, Suzzallo Library, FM-25, Map Collection, Seattle, WA 98195.

Clarke, Keith C. *Analytical and Computer Cartography.* Englewood Cliffs, N.J.: Prentice Hall, 1990. An informative introduction to the principles of cartographic data structures and algorithms for readers with a fuller interest in computer-assisted mapping.

Cuff, David J., and Mark T. Mattson. *Thematic Maps: Their Design and Production.* New York: Routledge, 1983. A concise, well-illustrated undergraduate text focused on the design and execution of statistical maps, locator maps, and distribution maps. Especially useful for describing the production of maps with a single color added to highlight specific features or enhance the design.

Dent, Borden D. *Cartography: Thematic Map Design.* 3d ed. Dubuque, Iowa: William C. Brown, 1993. A lucid and comprehensive introduction to map design. Especially useful for its treatment of color, human factors in map design, photomechanical reproduction techniques, and traditional cartographic drafting.

Greenhood, David. *Mapping.* Rev. ed. Chicago: University of Chicago Press, 1964. This well-written, classic introduction to surveying and mapping concludes with practical advice on how to compile and draw maps using inexpensive equipment.

Holmes, Nigel. *Pictorial Maps.* New York: Watson-Guptill, 1991. This rich, varied, and carefully reproduced collection of straightforward, dramatic maps may be useful as a source of ideas.

Keates, J. S. *Cartographic Design and Production.* 2d ed. London: Longman, 1989. An intermediate-level textbook describing the media and equipment of map production and their use for producing both black-and-white and color maps. Provides a useful overview of the process of producing complex multicolor maps, which can involve numerous feature and color separations and can require careful planning, as a result.

Makower, Joel, Cathryn Poff, and Laura Bergheim, eds. *The Map Catalog: Every Kind of Map and Chart on Earth and Even Some Above It.* Rev. ed. New York: Vintage, 1990. Source of titles, addresses, and an amazing amount of other information about maps, atlases, and other cartographic products potentially useful for research and map compilation.

Monkhouse, F. J., and H. R. Wilkinson. *Maps and Diagrams: Their Compilation and Construction.* 3d ed. London: Methuen, 1971. Although dated by its pre-1960 approach to pen-and-ink drafting, this classic British textbook is a source of useful ideas for graphing and mapping geographic data.

Monmonier, Mark. *How to Lie with Maps*. Chicago: University of Chicago Press, 1991. Introduces cartographic generalization and promotes a healthy skepticism about cartographic representation. Especially useful for its discussion of the effects on data maps of classification, symbolization, and aggregation.

Muehrcke, Phillip C. *Map Use: Reading, Analysis, and Interpretation*. 3d ed. Madison, Wis.: JP Publications, 1992. A popular college text on map reading, especially useful for its discussion of map analysis, map accuracy, and a variety of sources of maps and other cartographic data.

Robertson, Bruce. *How to Draw Charts and Diagrams*. Cincinnati: North Light Books, 1988. A well-illustrated introduction to traditional drafting media, as well as to graphic principles for designing straightforward, eye-catching data graphics. The chapter on maps includes a variety of useful examples.

Robinson, Arthur H., Randall D. Sale, Joel L. Morrison, and Phillip C. Muehrcke. *Elements of Cartography*. 5th ed. New York: John Wiley and Sons, 1984. *Elements* has been the bible of academic cartography since the early 1950s, and is widely respected as the richest, most detailed elementary cartographic text.

Snyder, John P., and Philip M. Voxland. *An Album of Map Projections*. U.S. Geological Survey Professional Paper no. 1453. Washington, D.C.: Government Printing Office, 1989. An excellent reference for advice and suggestions on choosing a map projection. Includes over 130 illustrations and catalogs the assets, deficiencies, common modifications, history, and usage of 90 basic projections.

Southworth, Michael, and Susan Southworth. *Maps: A Visual Survey and Design Guide*. Boston: Little, Brown, 1982. A fascinating source of examples of the effective use and design of maps in both color and black-and-white.

Szegö, Janos. *Human Cartography: Mapping the World of Man*. Stockholm: Swedish Council for Building Research, 1987. A fascinating discussion of map-design theory and source of ideas for maps.

Tufte, Edward R. *Envisioning Information*. Chesire Conn.: Graphics Press, 1990.

———. *Visual Display of Quantitative Information*. Cheshire, Conn.: Graphics Press, 1983. Tufte's two books on information graphics contain a variety of excellently reproduced examples of highly effective data graphics, including maps, and offer numerous insightful suggestions on the analysis and display of data.

Notes

CHAPTER 2

1. Conformality refers in theory to the preservation of the shapes of infinitesimally small circles. Although a conformal projection will in principle distort a circle of finite radius on the earth, circles that are as much as several miles in radius still appear on large-scale maps to be perfect circles. The distortion resulting from projection alone of traffic circles, circular buildings, crescent-shaped objects, and other local features is undetectable. Although technically incorrect, the assertion that conformal projections preserve shape on large-scale maps is a generally safe interpretation of conformality for maps at scales of 1:25,000 and larger.

2. The National Geographic Society, a respected, highly conservative organization, made the decision to adopt the Robinson projection after considerable deliberation. Once made, the decision attracted the attention of the media. See, for example, John B. Garver, Jr., "New Perspective on the World," *National Geographic* 174, no. 6 (December 1988), 910–13; and Bruce Van Voorst, "The New Shape of the World: A Pioneering Map Presents a Fresh View of Reality," *Time* 132, no. 19 (7 November 1988), 127.

3. For a discussion of the Tissot's indicatrix and its use, see Piotr H. Laskowski, "The Traditional and Modern Look at Tissot's Indicatrix," *American Cartographer* 16, no. 2 (April 1989), 123–33.

4. For an early presentation of this projection, see J. Paul Goode, "The Homolosine Projection: A New Device for Portraying the Earth's Surface Entire," *Annals of the Association of American Geographers* 15, no. 3 (September 1925), 119–25. By interrupting the earth over land, a variant of Goode's projection serves the needs of map authors concerned with oceanography, naval strategy, and marine affairs.

5. Making the projection secant, rather than tangent, yields two

267

lines of contact and a wider zone of low-to-moderate distortion. For a secant projection, the two standard lines are not meridians but small circles spaced a uniform distance to the east and west of a central meridian. For large-scale topographic maps, many national mapping agencies, including the U.S. Geological Survey, now use the *transverse Mercator* projection, a conformal representation that can be either secant or tangent. Until the 1950s, the Geological Survey used a *polyconic* projection for its large-scale topographic maps. Neither conformal nor equal-area, a polyconic provides true scale along all parallels but only along its central meridian. See John P. Snyder, *Map Projections Used by the U.S. Geological Survey*, U. S. Geological Survey Bulletin 1532 (Washington, D.C., 1982), 56–57, 126–27.

6. Richard Edes Harrison, who drew maps for *Fortune* and *Time* during and after World War II, used a variety of oblique azimuthal projections to capture some illuminating views of geopolitical relationships. According to political scientist Alan Henrikson, these maps promoted an "air-age globalism" that dramatized how the United States was drawn into the war and how the Soviet Union posed a threat to American security during the Cold War that followed. See Alan K. Henrikson, "The Map as an 'Idea': The Role of Cartographic Imagery during the Second World War," *American Cartographer* 2, no. 1 (April 1974), 19–53.

7. For a brief discussion of the correct use of the Mercator chart, the dangers of its misuse, and its abuse by political propagandists, see Mark Monmonier, *How to Lie with Maps* (Chicago: University of Chicago Press, 1991), 14–16, 94–96 .

8. For a discussion of tilted oblique-perspective views of the earth, see John P. Snyder, "The Perspective Map Projection of the Earth," *American Cartographer* 8, no. 2 (October 1981), 149–60.

CHAPTER 3

1. Bertin's most influential work is his *Sémiologie Graphique*, first published in 1967. Even before an English translation appeared in 1983, North American and British cartographers, as well as graphic theorists outside cartography, recognized the significance and utility of Bertin's attempt to develop an organized set of principles for graphic design based in semiotic logic. See Jacques Bertin, *Semiology of Graphics: Diagrams, Networks, Maps*, trans. William J. Berg (Madison: University of Wisconsin Press, 1983).

2. A reversed metaphor might logically be constructed as "the brighter the more intense, the lighter the less intense." For graphics printed on paper with ink or toner, a notion of "the greater the more" is intuitive, because it associates "more" with the deliberate act of adding more ink or toner to produce a darker symbol. We can call this a "printing metaphor." The reversed construction would be equally valid if the

background of the graphic was comparatively dark and what was added was either a reflective ink or light, as on a video monitor. But because electronic publishing systems often generate a light-toned background so that text and graphic marks stand out as dark, we cannot label this notion a "video (or CRT) metaphor."

3. "Percentages" requires an important caveat. The map author must make a clear distinction between percentage data that represent a place's proportionate share of a total for the region and percentage data that represent the proportion of each place's own total with a particular attribute, such as urban residence. The former measure is really a transformation obtained by dividing every count by the regional sum, whereas the latter is a true intensity measure. Thus, a map showing each county's percentage share of the overall state or provincial population would be similar in its geographic pattern to a map showing the sizes of urban populations by county. In contrast, a county-unit map of the "urban percentage of the population" would reflect not the numbers of people living in urbanized places but the intensity with which people have chosen (or been induced) to live in an urban environment.

4. Cartographers sometimes produce three-dimensional, or volumetric, symbols, such as the plaster or epoxy terrain representations used in museum exhibits or in landscape architects' project models. These are true volumetric symbols. Development of efficient, high-speed computer graphics technology is fostering an increased use of dynamic three-dimensional maps that the viewer can interact with through rotation, zoom-in, fly-over, and other manipulations. These dynamic symbols are also volumetric symbols and thus beyond the scope of this book; they contrast markedly with static symbols that employ perspective foreshortening, shadows, and other depth clues to induce a visual impression of three dimensions. I treat the latter merely as dramatically enhanced versions of point, line, and area symbols. For the next decade or so, true volumetric symbols will probably remain beyond the reach and the grasp of most humanists and social scientists.

5. Experimental studies in psychophysics have long recognized the tendency of subjects to underestimate the values of larger stimuli. S. S. Stevens recognized a general log-linear stimulus-response relationship, which he described with the power function now called Stevens' Law. See, for example, Gosta Ekman, Ralf Lindman, and W. William-Olson, "A Psychophysical Study of Cartographic Symbols," *Perceptual and Motor Skills* 13, no. 3 (December 1961), 355–68; and S. S. Stevens, "On the Psychophysical Law," *Psychological Review* 64, no. 3 (May 1957), 153–81. Among the first studies to demonstrate the magnitude of this effect for cartographic symbols was James John Flannery's doctoral dissertation, defended in 1956 but not published as a journal article until 1971. See James John Flannery, "The Relative Effectiveness of Some Common Graduated Point Symbols in the Presentation of Quantitative

Data," *Canadian Cartographer* 8, no. 2 (December 1971), 96–109; and Kang-Tsung Chang, "Visual Estimation of Graduated Circles," *Canadian Cartographer* 14, no. 2 (December 1977), 130–38. Because point symbols scaled by area or volume tend to understate the larger of two values, precise pair comparison might require either linear scaling or *apparent-value rescaling*, a technique that attempts to trick the map viewer by representing large values with larger, deliberately exaggerated symbols to compensate for visual underestimation. Although cartographic textbooks commonly describe the mechanics of apparent-value rescaling, the technique is not widely used, probably because most map authors recognize the hopelessness of using a correction based on average responses to deal with the enormous variations among individual map viewers. See, for example, Borden D. Dent, *Cartography: Thematic Map Design*, 2d ed. (Dubuque, Ia.: William C. Brown, 1990), 201–3.

6. For discussion of the role of representative anchor stimuli, see Carleton W. Cox, "Anchor Effects and the Estimation of Graduated Circles and Squares," *American Cartographer* 3, no. 1 (April 1976), 65–74.

7. For the rationale underlying frame-rectangle point symbols, see William S. Cleveland and Robert McGill, "Graphical Perception: Theory, Experimentation, and Application to the Development of Graphical Methods," *Journal of the American Statistical Association* 79, no. 387 (September 1984), 531–54. For a comparative evaluation of the framed rectangle's effectiveness as a cartographic symbol, see Richard Dunn, "Framed Rectangle Charts or Statistical Maps with Shading: An Experiment in Graphical Perception," *American Statistician* 42, no. 2 (May 1988), 123–29.

8. For an insightful essay on the iconology of map symbols, see Denis Wood and John Fels, "Designs on Signs: Myth and Meaning in Maps," *Cartographica* 23, no. 3 (Autumn 1986), 54–103.

9. For an especially comprehensive treatment of flow maps in a cartographic textbook, see Borden D. Dent, *Cartography: Thematic Map Design*, 2d ed. (Dubuque, Ia.: William C. Brown, 1990), 259–78.

10. For a rich variety of isoline maps, see Ronald Abler, John S. Adams, and Peter Gould, *Spatial Organization: The Geographer's View of the World* (Englewood Cliffs, N.J.: Prentice-Hall, 1971). This basic geography textbook was well ahead of its time, sold poorly, and is now out-of-print. But even though some of its examples might seem out-of-date, *Spatial Organization* offers a solid overture to theoretical human geography.

11. Area symbols can indeed vary in size, as do the area cartograms discussed in chapter 6. But contiguous-area cartograms are transformations of space akin to map projections, and noncontiguous-area cartograms employ a graduated point symbol that reflects the shape of the areal unit. Area symbols might also vary according to the orientation of

a coarse pattern of similarly oriented lines or arrowheads, but such a pattern is much less effective than a single arrow for representing direction.

12. Cartographic researchers have devoted considerable effort to fine-tuning an equal-value or apparent-value gray scale in which each successive pair of graytones in a graduated series has an equal perceived difference. See, for example, Carleton W. Cox, "The Effects of Background on the Equal Value Gray Scale," *Cartographica* 17, no. 1 (Spring 1980), 53–71; and A. Jon Kimerling, "The Comparison of Equal-Value Gray Scales," *American Cartographer* 9, no. 1 (April 1985), 132–42. This goal has merit if ink spread is both predictable and carefully controlled. Yet a simulation model suggests that because the effect of ink spread on dot screens with opaque dots (representing values under 50-percent black) is different from its effect on dot screens with clear dots (representing values over 50-percent black), the printed version of a 45-percent black-dot symbol might actually be darker than the printed version of a 55-percent black-dot symbol. See Mark Monmonier, "The Hopeless Pursuit of Purification in Cartographic Communication: A Comparison of Graphic-Arts and Perceptual Distortions of Graytone Symbols," *Cartographica* 17, no. 1 (Spring 1980), 24–39.

13. The use of color is beyond the scope of this essay, because of the generally prohibitive cost of reproducing polychrome maps in print. Color might be conveniently available for maps and other graphics produced as 35mm color transparencies with an electronic slide generator. The map author might either find a slide generator on campus in a central computer graphics facility or send files by mail to an electronic slide service. The slide service might even provide free software to help the user convert graphics composed with drawing software to the format required for its slide generator. Often a map prepared for print publication can be suitably modified by adding color, dropping some labels, and making the remaining labels larger and more legible on the projection screen. If used tastefully and logically, color can be a valuable source of visual contrast for representing differences among areas. For mapping quantitative data, though, color should neither be the dominant visual variable and nor conflict with the value contrasts (lighter-darker contrast) representing quantitative differences among data values. For further reading on the proper and effective use of color on maps, see Judy M. Olson, "Color and the Computer in Cartography," in H. John Durrett, ed., *Color and the Computer* (Orlando, Fla.: Academic Press, 1987), 205–19; and Edward R. Tufte, *Envisioning Information* (Cheshire, Conn.: Graphics Press, 1990), 53–65, 81–96.

14. See, for example, Mark Monmonier, *Maps with the News: The Development of American Journalistic Cartography* (Chicago: University of Chicago Press, 1989), 2–3, 180–86.

15. For an account of the fire, with maps and photographs describing the extent of destruction and the fire's effects on the Chicago landscape, see Harold M. Mayer and Richard C. Wade, *Chicago: Growth of a Metropolis* (Chicago: University of Chicago Press, 1969), 105–10 .

16. A "minor civil division" is a political subdivision of a county, such as a township, city, or incorporated village.

17. For a fascinating collection of a variety of Zelinsky's maps and similar work, see John F. Rooney, Jr., Wilbur Zelinsky, and Dean R. Louder, eds., *This Remarkable Continent: An Atlas of the United States and Canadian Society and Cultures* (College Station: Texas A & M University Press, 1982).

18. The reader interested in countable symbols developed using the internal elements of point, line, and area symbols should see Roberto Bachi, *Graphical Rational Patterns: A New Approach to Graphical Presentation of Statistics* (Jerusalem: Israeli Universities Press, 1968). Although "rational," as his title implies, Bachi's complex graphic code is alien and confusing to most map viewers.

19. James W. Cerny, "Joyce's Mental Map," *James Joyce Quarterly* 9, no. 2 (Winter 1971), 218–24.

CHAPTER 4

1. Maps can decorate books in a variety of ways. For a hardbound book, an interesting portion of a particularly attractive or fascinating map might serve as an endpaper. Maps can also enhance a book's design as artwork on the cover. Book covers are important point-of-sale ads for the publisher, and a carefully selected map, integrated creatively with the title, can catch the prospective buyer's eye and communicate at a glance the content and value of a book that focuses on a specific place, region, or geographic phenomenon.

2. Mark Monmonier, "Map-Text Coordination in Geographic Writing," *Professional Geographer* 33, no. 4 (November 1981), 406–12. This essay also argues for close physical proximity of the map and its associated text in the layout of a book or journal.

3. *The Chicago Manual of Style*, 14th ed. (Chicago: University of Chicago Press, 1993), 11.7. My original reference was to the 12th edition (1969), p. 258, which supported my position as explicitly as the current 14th edition. In its comparatively terse discussion of references to illustrations from the text, the 11th edition (1949), p. 52, neither included an example with the illustration as the subject of a sentence nor proscribed such usage.

4. Ibid., 11.6.

5. Doctoral candidates who aspire to publish their dissertations as books should be especially wary of the difference in page size between dissertations and books. Type and symbols easily decoded on a complex

illustration prepared for the letter-size page of a thesis might be incomprehensible if reduced to fit the substantially smaller format of a university press book. Maps with type and symbols that might look a bit large or exaggerated on letter-size pages in the dissertation will not need to be subdivided and redrawn later for publication. They will also accommodate the reduced size of bound copies produced from microfilm, as well as the diminished visual acuity of senior faculty members.

6. If the publication addresses the general public, a supplementary note somewhere on the map, in the caption, or in the book or article might call attention to or even provide the precise yet somewhat complicated definition of "urban area" used by the Bureau of the Census. For an audience of social scientists competent in demography, this information should be unnecessary.

7. In *Semiology of Graphics*, Jacques Bertin discusses briefly the role of map titles but says little else about type, which is not a part of his graphic system.

8. Colored type can expedite the search for a label if the viewer knows what color to look for. But poor contrast with the background can make colored type difficult to read. For discussion of legibility and the use of colored type, see Barbara S. Bartz, "Experimental Use of the Search Task in an Analysis of Type Legibility in Cartography," *Cartographic Journal* 7, no. 2 (December 1970), 103-12; and Richard J. Phillips, Liza Noyes, and R. J. Audley, "The Legibility of Type on Maps," *Ergonomics* 20, no. 6 (November 1977), 671–82.

9. Labels rendered in gray, rather than black, can be large yet visually recessive. Type produced in gray using narrow or medium typefaces tends to look jagged and "break up." Gray type works best with typefaces large or bold enough to provide sufficient contrast with the background.

10. For a summary of experimental studies in the legibility of cartographic and other lettering, see Barbara S. Bartz, "An Analysis of the Typographic Legibility Literature," *Cartographic Journal* 7, no. 1 (June 1970), 10–16.

11. Both designers and subject-testing studies suggest that at least a 2-point difference in font size is needed to make labels clearly distinct. See, for example, Barbara Gimla Shortridge, "Map Reader Discrimination of Lettering Size," *American Cartographer* 6, no. 1 (April 1979), 13–20.

12. Sans serif typefaces are seldom used for the so-called bodytype, in which the text of a book or journal is set. My preference for sans serif for street names set in tiny, narrow type reflects a worry that fragile, thin serifs on very small type might not survive image transfer during printing or plate making. In particular, photoreduction might either partly erase some serifs or fill in the corners where serifs meet main strokes.

13. For discussion of preferred positions for cartographic labels, see

Pinhas Yoeli, "The Logic of Automatic Map Lettering," *Cartographic Journal* 9, no. 2 (December 1972), 99–108.

14. Automating label placement has been a thorny problem in computer-assisted cartography. For examples of various trial-and-error strategies used to automate label placement, see Lee R. Ebinger and Ann M. Goulette, "Noninteractive Automated Names Placement for the 1990 Decennial Census," *Cartography and Geographic Information Systems* 17, no. 1 (January 1991), 69–78; Stephen A. Hirsch, "An Algorithm for Automatic Name Placement around Point Data," *American Cartographer* 9, no. 1 (April 1982), 5–17; and Steven Zoraster, "Integer Programming Applied to the Map Label Placement Problem," *Cartographica* 23, no. 3 (Autumn 1986), 16–27.

15. For a comprehensive illustrated essay on positioning placename labels, see Eduard Imhof, "Positioning Names on Maps," *American Cartographer* 2, no. 2 (October 1975), 128–44.

16. A human-factors study concluded that search for names in lowercase type was easier than search for names in uppercase type of the same point size; see Richard J. Phillips, Elizabeth Noyes, and R. J. Audley, "Searching for Names on Maps," *Cartographic Journal* 15, no. 2 (December 1978), 72–77.

17. For examples of cartographic problems caused by different systems of transliteration, see Joseph R. Morgan, "Southeast Asian Place Names: Some Lessons Learned," *American Cartographer* 11, no. 1 (April 1984), 5–13.

18. Hans Kurath and others, *Linguistic Atlas of New England* (Providence, R.I.: Brown University, 1939–43). For a description and interpretation of the *Atlas*, see Hans Kurath and others, *Handbook of the Linguistic Geography of New England* (Providence: Brown University, 1939). Also see Hans Kurath, *A Word Geography of the Eastern United States*, Studies in American English no. 1 (Ann Arbor: University of Michigan Press, 1949). For a review of the *Atlas*, see William Cabel Greet, "Review of the Atlas Handbook," *American Speech* 15, no. 2 (April 1940), 185–89; also see Raven I. McDavid, "Dialect Geography and Social Science Problems," *Social Forces* 25, no. 2 (December 1946), 168–72.

CHAPTER 5

1. For concise histories of county atlas publishing in the United States, see Michael P. Conzen, "The County Land-Ownership Map in America: Its Commercial Development and Social Transformation, 1814–1939," *Imago Mundi* 36 (1984), 9–31; and Norman J. W. Thrower, "The County Atlas in the United States," *Surveying and Mapping* 21, no. 3 (September 1961), 365–73. Maps and pictorial engravings included in county atlases seem generally reliable and useful as historical documents.

2. For a discussion of American fire-insurance maps, see Walter W. Ristow, "United States Fire Insurance and Underwriters Maps, 1852–1968," *Quarterly Journal of the Library of Congress* 25, no. 3 (July 1968), 194–218. For discussion of early aerial surveying for urban areas, see Dino Brugioni, "Aerial Photography: A Challenge and a Commitment," *Perspectives* (American Historical Association Newsletter) 23, no. 8 (November 1985), 6, 10–11; Teodor J. Blachut and Rudulf Burkhardt, *Historical Development of Photogrammetric Methods and Instruments* (Falls Church, Va.: American Society for Photogrammetry and Remote Sensing, 1989); Nelson F. Pitts, Jr., "What Aerial Mapping Has Done for Syracuse," *American City* 37, no 3 (September 1927), 354–56; and Chester C. Slama and others, *Manual of Photogrammetry*, 4th ed. (Falls Church, Va.: American Society of Photogrammetry, 1980), 2–29.

3. For an overview of federal cartographic activities and a concise description of the National Map Library Depository Program, see Gary W. North, "Maps for the Nation: The Current Federal Mapping Establishment," *Government Publications Review* 10, no. 4 (July–August 1983), 345–60. Also see Mary Lynette Larsgaard, *Map Librarianship: An Introduction* (Littleton, Colo.: Libraries Unlimited, 1987), 61–74.

4. The Geological Survey has published a number of useful guides to its products. Particularly helpful are Morris M. Thompson, *Maps for America: Cartographic Products of the U.S. Geological Survey and Others*, 3d ed. (Washington, D.C.: Government Printing Office, 1987); and U.S. Geological Survey, National Mapping Program, *Map Data Catalog* (Washington, D.C.: Government Printing Office, 1984). In addition, the Geological Survey has computer-readable cartographic data sets and other cartographic products, as well as a number of free booklets and posters describing topographic maps, map projections, and aerial photographs.

5. The Central Intelligence Agency's cartographic section does excellent work, and however ironic it might seem, makes its maps, atlases, and cartographic data files available to the general public at a modest cost. Perhaps some day the CIA will even allow staff members attending professional meetings to wear name tags that identify their employer more specifically than the telltale generic "United States Government."

6. Order unclassified CIA maps from the Superintendent of Documents, Government Printing Office, Washington DC 20402, or the National Technical Information Service, Springfield VA 22161; consult current catalogs in the map library or the library's government documents section for product numbers and prices. Request information from the Geological Survey by calling the National Cartographic Information Center at 1-800-USA-MAPS, or by writing to the National Cartographic Information Center, 507 National Center, U.S. Geological Survey, 12201 Sunrise Valley Drive, Reston VA 22092.

7. See Rupert B. Southard, "The Development of U.S. National Mapping Policy," *American Cartographer* 10, no. 1 (April 1983), 5–15.

8. The 1:25,000 scale reflects the largely abortive move during the 1970s toward scales consistent with the metric system. The traditional scale for the 7.5-minute series, 1:24,000, reflects so-called British units (1 inch on the map represents 2,000 feet on the ground). According to a metric policy adopted in 1977, states could choose between 1:24,000 and the nearest "metric" scale, 1:25,000, as well as between feet and meters for the intervals between elevation contours. Although a few states have new or revised large-scale maps published at 1:25,000, most preferred to retain 1:24,000 for the sale of tradition and consistency.

9. For an excellent overview of detailed base maps and other cartographic products available for Canada and the United Kingdom, see N. L. Nicholson and L. M. Sebert, *The Maps of Canada: A Guide to Official Canadian Maps, Charts, Atlases and Gazetteers* (Folkestone, Kent: William Dawson and Sons; Hamden, Conn.: Archon Books, distributed by Shoe String Press, 1981); and J. B. Harley, *Ordnance Survey Maps: A Descriptive Manual* (Southampton: Ordnance Survey, 1975).

10. For a thorough examination of cataloging and other challenges in managing a map collection, see Mary Lynette Larsgaard, *Map Librarianship: An Introduction* (Littleton, Colo.: Libraries Unlimited, 1987).

11. For discussion of the variety of atlases and atlas formats, see Mark Monmonier, "Trends in Atlas Development," *Cartographica* 18, no. 2 (Summer 1981), 187–213.

12. Libraries at universities have tended to place digital satellite imagery, census tapes, and other large machine-readable geographic data sets in the hands of computer services units. Change seems likely, though, as libraries adopt a broader view of their mission as information providers. For an insightful discussion of the evolution and growth of map libraries, see Walter W. Ristow, *The Emergence of Maps in Libraries* (Hamden, Conn.: Linnet Books, Mansell Publishing, 1980).

13. John A. Wolter, Ronald E. Grim, and David K. Carrington, eds., *World Directory of Map Collections*, 2d ed. (Munich: K. G. Saur, 1986).

14. David K. Carrington and Richard W. Stephenson, eds., *Map Collections in the United States and Canada: A Directory*, 4th ed. (New York: Special Libraries Association, 1985); and David A. Cobb, ed., *Guide to U.S. Map Resources*, 2d ed. (Chicago: American Library Association, 1990).

15. Ian Watt, *A Directory of U.K. Map Collections*, 2d ed. (Kingston-upon-Thames, Surrey: British Cartographic Society, 1985).

16. Ralph E. Ehrenberg, *Scholar's Guide to Washington, D.C., for Cartography and Remote Sensing* (Washington, D.C.: Smithsonian Institution Press, 1987).

17. Jeffrey A. Kroessler, *A Guide to Historical Map Resources for Greater New York* (Chicago: Speculum Orbis Press, 1988).

18. These catalogs can be useful sources of specific information for research proposals. The National Endowment for the Humanities offers small grants through its Travel to Collections program. The Hermon Dunlap Smith Center for the History of Cartography at the Newberry Library, in Chicago, and the John Carter Brown Library at Brown University, in Providence, Rhode Island, both have fellowship programs that provide a small stipend, as well as desk space and assistance in using the collection.

19. The British Museum published the British Library's map catalog. The G. K. Hall Company of Boston, a prominent publisher of facsimile catalogs, has published, among others, the *Catalog of the National Map Collection of the Public Archives of Canada*; *Dictionary Catalog of the Map Division of the New York Public Library*; *Research Catalogue of the American Geographical Society* (originally in New York City but at the University of Wisconsin at Milwaukee since the late 1970s); *Catalog of Manuscript and Printed Maps in the Bancroft Library* (University of California, Berkeley); and *Research Catalog of Maps of America to 1860 in the William L. Clements Library* (University of Michigan).

20. C. R. Perkins and R. B. Parry, eds., *Information Sources in Cartography* (London: Bowker-Saur, 1990).

21. R. B. Parry and C. R. Perkins, eds., *World Mapping Today* (London: Butterworths, 1987).

22. Rolf Bohme, *Inventory of World Topographic Mapping* (London: Elsevier Applied Science Publishers, 1989–). The *Inventory* is organized by regional volumes; volume 1, for instance, includes Western Europe, North America, and Australasia.

23. K. F. Kister, *Kister's Atlas Buying Guide: General English-Language World Atlases Available in North America* (Phoenix, Ariz.: Oryz Press, 1984); W. Stams, *National and Regional Atlases* (Enschede, Netherlands: International Cartographic Association, 1984); and Kenneth L. Winch, *International Maps and Atlases in Print*, 2d ed. (London: Bowker, 1976).

24. *Bibliographic Guide to Maps and Atlases* (Boston: G. K. Hall, 1979–).

25. Gloria-Gilda Deak, *Picturing America, 1497–1899: Prints, Maps, and Drawings Bearing on the New World Discoveries and on the Development of the Territory That Is Now the United States* (Princeton, N.J.: Princeton University Press, 1988); Bernard Romans, *Georgia at the Time of the Ratification of the Constitution, from Original Maps in the Library of Congress at Washington* (Washington, D.C.: U.S. Geological Survey, 1937); John R. Sellers, *Maps and Charts of North America and the West Indies, 1750–1789: A Guide to the Collections in the Library of Congress* (Washington, D.C.: Library of Congress, 1981); and James Clements Wheat, *Maps and Charts Published in America*

before 1800: A Bibliography (New Haven, Conn.: Yale University Press, 1969).

26. John R. Hebert and Patrick E. Dempsey, *Panoramic Maps of Cities in the United States and Canada: A Checklist of Maps in the Collections of the Library of Congress, Geography and Map Division,* 2d ed. (Washington, D.C.: Library of Congress, 1984); Library of Congress, Geography and Map Division, Reference and Bibliography Section, *Fire Insurance Maps in the Library of Congress: Plans of North American Cities and Towns Produced by the Sanborn Map Company: A Checklist* (Washington, D.C.: Library of Congress, 1981); Richard W. Stephenson, *Civil War Maps: An Annotated List of Maps and Atlases in the Library of Congress,* 2d ed. (Washington, D.C.: Library of Congress, 1989); and Richard W. Stephenson, *Land Ownership Maps: A Checklist of Nineteenth Century United States County Maps in the Library of Congress* (Washington, D.C.: Government Printing Office, 1967).

27. Eran Laor and Shoshanna Klein, *Maps of the Holy Land: Cartobibliography of Printed Maps, 1475–1900* (New York: A. R. Liss; Amsterdam: Meridian, 1986).

28. American Geographical Society, Map Department, *Index to Maps in Books and Periodicals* (Boston: G. K. Hall, 1968; supplements, 1971, 1976, 1986).

29. David C. Jolly, *Maps of America in Periodicals before 1800* (Brookline, Mass., 1989).

30. J. B. Harley and David Woodward, eds., *History of Cartography* (Chicago: University of Chicago Press, 1987–); and U.S. Library of Congress, Geography and Map Division, *Bibliography of Cartography* (Boston: G. K. Hall, 1973; supplement, 1980).

31. The number of words an author may quote without obtaining permission seems to vary with the length of the work and the prominence of the author. Although quoting one line of a Robert Frost poem without permission might be a cause for legal action, or at least a nasty letter, one could probably quote 100 words by Stephen King or 500 (or even 1,000) words by Mark Monmonier with impunity. Prentice-Hall, for example, has a guideline of 250 words from any single work but lists a number of exceptions, such as poems, dramatic scripts, and unpublished letters; see *Prentice-Hall Author's Guide,* 5th ed. (Englewood Cliffs, N.J.: Prentice-Hall, 1978), 8–9. As pragmatic guidelines devised by publishers to reduce the workload of permissions departments, these rules have no legal force, because the law does not state specifically how much one may copy under the doctrine of fair use. See *The Chicago Manual of Style,* 14th ed. (Chicago: University of Chicago Press, 1993), 4.51–4.58.

32. A schedule of per-copy fees is graduated according to the area of mapping to be reproduced or used and whether the reproduction is fac-

simile or compiled. Although royalties are normally waived for "non-profit-making publications of academic research," the Ordnance Survey requires that the scholar apply for permission and include a notice recognizing the Crown copyright. There is also no charge when Ordnance Survey maps are used to compile a redrawn map at a scale smaller than 1:1,000,000. See Ordnance Survey, "Copyright–Publishing," *OS Information Leaflet* no. 23 (January 1985). Different rules apply to copyright-protected maps and atlases published by other public agencies in Great Britain.

33. For a concise examination of copyright laws in Great Britain and the United States, see Christopher Scarles, *Copyright* (Cambridge: Cambridge University Press, 1980). For an intriguing examination of the attitude of Congress and the courts toward maps and the attitude of map publishers toward copyright registration, see James W. Cerny, "Awareness of Maps as Objects for Copyright," *American Cartographer* 5, no. 1 (April 1978), 45–56.

34. Because the law also grants the copyright owner the right to control the means and manner in which a work is reproduced, altering a work not in the public domain without obtaining permission might infringe the owner's copyright. Actionable alterations include (but are not limited to) adding or eliminating features in a manner that falsely suggests incompetence or dishonesty on the part of the original map author.

35. For discussion of trap streets, see William Alexander Miller, "The Copyright of a Map or Chart," *National Geographic Magazine* 13, no. 12 (December 1902), 437–43; and Mark Monmonier, *How to Lie with Maps* (Chicago: University of Chicago Press, 1991), 47–49.

36. Considerable complexity used to surround unpublished material and the fact or instant of "publication." Establishing that the copyright notice was omitted or incorrect when the map was first published was not easy. The absence of a copyright notice such as "© Rand McNally 1964" did not necessarily mean the map was in the public domain, because the version an author wanted to copy might itself have been an unauthorized, illegal reproduction of a validly copyrighted map. Moreover, until 1978, common law protection emanated from the states, not the federal government. In the Copyright Act of 1978, Congress replaced the somewhat vague notion of publication with a simpler threshold requirement of fixation. Federal copyright now protects all unpublished works, if they are fixed in tangible form. For further discussion of this important shift in copyright legislation, see William S. Strong, *The Copyright Book: A Practical Guide*, 3d ed. (Cambridge, Mass.: M.I.T. Press, 1990), 2–3, 69–70, 160–64.

37. For jointly written works, the term of copyright depends upon the date of death of the last surviving coauthor; for works produced "for hire," such as most commercial atlases and street maps, protection lasts either 75 years from the first publication of the work or 100 years from

its creation, whichever term would expire first. The scholar eager to avoid infringing someone's copyright need feel little uncertainty about the duration of copyright for post-1978 material. If properly registered and suitably eligible, these works almost always are still covered, and the author who wants to reproduce them must seek permission. See, for example, ibid., 46–49.

38. Ibid., 169–70, 177–78; and *The Chicago Manual of Style*, 4.8.

39. A research library with a good collection of bibliographic references should have the *Catalog of Copyright Entries*, published by the Library of Congress. In addition, the Copyright Office can determine a work's status by conducting a copyright search for a nominal hourly fee.

40. European copyrights have had a 50-year-plus-life duration longer than American copyrights. For information about international copyright protection, consult the Copyright Office, 1921 Jefferson Davis Highway, Arlington VA 22201, or see Strong, *The Copyright Book*, 195–99.

41. See Mark Monmonier, *Maps with the News: The Development of American Journalistic Cartography* (Chicago: University of Chicago Press, 1989), 37. Although I was unable to find a copy of the original drawing as published in the Boston *Gazette* for March 26, 1812, I located a clearly detailed example in James Parton's *Caricature and Other Comic Art* (New York: Harper and Brothers, 1877), 316.

42. For general discussion of fair use, see John Shelton Lawrence and Bernard Timberg, *Fair Use and Free Inquiry* (Norwood, N.J.: ABLEX, 1980); and Strong, *The Copyright Book*, 129–69.

43. *The Chicago Manual of Style*, 4.51.

44. See, for example, R. S. Talab, *Commonsense Copyright: A Guide to the New Technologies* (Jefferson, N.C.: McFarland, 1986), 19–22.

45. Attorneys for map publishers seem likely to disagree. See, for example, James M. Votava, "Map Copyrights," *Surveying and Mapping* 7, nos. 3, 4 (1947), 216–18. Votava was general counsel for Rand McNally and Company. However, suits charging infringement have often failed, except for wholesale copying, because the courts have often taken a dim view of cartographic creativity; see John F. Whicher, "Originality, Cartography, and Copyright," *New York University Law Review* 38, no. 2 (April 1963), 280–300.

46. *Asia Today: An Atlas of Reproducible Pages* (Wellesley, Mass.: World Eagle, 1988). World Eagle has also published similar atlases for Africa, Europe, and Latin America.

47. I know of no case involving critical review or scholarship as the defense, successful or otherwise, for the unauthorized reproduction of a map or an excerpt therefrom, although there are numerous parallel cases in the visual arts. Court decisions have tended to support unauthorized copying of visual material to foster scholarship. But litigation or

the threat of litigation seems to have had a chilling effect on such reproduction of tobacco advertising and excerpts of films and videos; see David A. Kaplan and Debra Rosenberg, "They Want Their MTV Back," *Newsweek* 117, no. 20 (20 May 1991), 68; and Brian S. O'Malley, "Fair Use and Audiovisual Criticism," *Intellectual Property Law Review* 16 (1984), 447–71.

48. William F. Patry, *The Fair Use Privilege in Copyright Law* (Washington, D.C.: Bureau of National Affairs, 1985), 402–4.

49. For discussion of copyright law and contemporary photography, see Robert M. Cavallo and Stuart Kahan, *Photography: What's the Law?* (New York: Crown Publishers, 1979).

50. For discussion of the role of copyright law in the operation of a museum, see Marie C. Malaro, *A Legal Primer on Managing Museum Collections* (Washington, D.C.: Smithsonian Institution Press, 1985), 113–22.

51. For further discussion of the content and form of permission letters, see *The Chicago Manual of Style*, 4.63, 4.66.

52. The compilation-source map might also be used to prepare a reliability diagram that describes how relative accuracy varies across the compiled map. When a complex compiled map is crucial to the author's thesis, the book or article should include a reliability diagram as a normal and necessary part of scholarly documentation. See Arthur H. Robinson and others, *Elements of Cartography*, 5th ed. (New York: John Wiley and Sons, 1984), 134.

53. For discussion of various survey techniques and how they affect the accuracy of a source, see T. J. Blachut, A. Chrzanowski, and J. H. Saastamoinen, *Urban Surveying and Mapping* (New York: Springer-Verlag, 1979); and John Wright, *Ground and Air Survey for Field Scientists* (Oxford: Clarendon Press, 1982), esp. 77–99.

54. For useful advice on evaluating cartographic sources, see M. J. Blakemore and J. B. Harley, *Concepts in the History of Cartography: A Review and Perspective*, *Cartographica* monograph 26 (also vol. 17, no. 4), (1980), 54–75; and J. B. Harley, *Maps for the Local Historian: A Guide to the British Sources* (London: Bedford Square Press and National Council of Social Service, 1972).

55. Gerald R. Crone, "New Light on the Hereford Map," *Geographical Journal* 131, pt. 4 (December 1965), 447–62. For an insightful examination of the reliability of medieval small-scale maps, see David Woodward, "Reality, Symbolism, Time, and Space in Medieval World Maps," *Annals of the Association of American Geographers* 75, no. 4 (December 1985), 510–21.

56. For discussion of provenance and other problems of electronic cartographic databases, see Mark Monmonier, *Technological Transition in Cartography* (Madison: University of Wisconsin Press, 1985), esp. 191–93.

57. See Derek Howse, *Greenwich Time and the Discovery of the Longitude* (Oxford: Oxford University Press, 1980), 138–55.

58. For information on techniques for identifying an unknown projection, converting this source projection's coordinates to latitude-longitude referencing, and then computing plane coordinates based on the projection of the compilation base, see J. P. Snyder, *Computer-Assisted Map Projection Research*, U.S. Geological Survey bulletin 1629 (1985), 1–54.

59. For discussion of the dimensional stability of paper maps, see D. H. Maling, *Measurements from Maps: Principles and Methods of Cartometry* (Oxford: Pergamon Press, 1989), 196–207; and William Ravenhill and Andrew Gilg, "The Accuracy of Early Maps: Towards a Computer Aided Method," *Cartographic Journal* 11, no. 1 (June 1974), 48–52.

60. For information about compiling maps from aerial photographs, see D. N. Riley, *Air Photography and Archeology* (Philadelphia: University of Pennsylvania Press, 1987), 60–77. One solution to the problem of radial displacement is the orthophotoquad map, available for some 7.5-minute quadrangles and based on *orthophotos*, which are air-photo images with radial displacement removed electronically. For information about Geological Survey orthophotoquad maps, see Thompson, *Maps for America*, 135–42.

61. Program and database software change too frequently and radically to warrant discussion here of specific products. *American Demographics*, *MacWorld*, *PC World*, and similar periodicals are useful sources of current information about mapping software.

62. For superb examples of the reproduction of color and black-and-white cartographic facsimiles and of the range of maps and aerial photographs available as historical sources, see David Buisseret, ed., *From Sea Charts to Satellite Images: Interpreting North American History through Maps* (Chicago: University of Chicago Press, 1990); and David Buisseret, *Historic Illinois from the Air* (Chicago: University of Chicago Press, 1990).

CHAPTER 6

1. Although the software manual encourages the user to move the legend, add a more suitable title, and possibly modify the categories, neither the manual nor the program warns the user that the choropleth map's five graytones and darker-more/lighter-less graphic coding obscure differences among counts and distort patterns in the data. Indeed, the manual proudly includes examples of choropleth maps of count data. See, for example, *Atlas MapMaker 4.5 User's Manual* (San Jose, Calif.: Strategic Mapping, 1990), 33–34, 43–45.

2. Figure 6.2 took about ten minutes longer to produce than Figure 6.1. I "digitized" visual centers for each of the twenty-one counties and

five positions for the map key, requested proportional point symbols, suppressed the area symbols in Figure 6.1, experimented briefly with circle sizes, and copied the resulting map into my drawing software for refinement. See the discussion of point symbols and digitizing in ibid., 199–203. An adequate description of the complicated process needed to make Figure 6.2 with version 4.5 of MapMaker would require several illustrations and several hundred words. The user needs to read the manual and work with the program. I am optimistic that future versions of the software will either include visual centers with the county boundary files or enhance the program's "Layers" menu with a function for computing area centroids. Of course, if the software required the user to designate each "category" variable as either a magnitude or an intensity measure, the default display could be customized to provide a more graphically logical "success experience" for users entering raw count data.

3. MapMaker apparently adds proportional-point symbols to the map in reverse alphabetical order. In the right-hand example in Figure 6.4, for instance, the circle for Bergen County covers part of the overlapping circle for Essex County, which partly blocks circles for Hudson, Morris, Passaic, and Union counties.

4. Mapping software may behave strangely when the average number of places per category is not an integer. For instance, in generating a six-category quantile map for New Jersey's twenty-one counties (shown in Figure 6.7), MapMaker violated the spirit of quantile classing by assigning four counties to each of the five lower "sextiles" and only one county to the highest category. Few if any human cartographers would do this. The sextile map in the bottom row reflects a more even allocation, with either three or four places in each category. I first assigned four counties each to the top and bottom categories and three counties each to the second and fifth categories, and then allocated three counties to the third category and four counties to the fourth, because this result was slightly more homogeneous than a four-and-three allocation would have been.

5. For further discussion of area cartograms, see Borden D. Dent, "Communication Aspects of Value-by-Area Cartograms," *American Cartographer* 2, no. 2 (October 1975), 154–68; Judy M. Olson, "Noncontiguous Area Cartograms," *Professional Geographer* 28, no. 4 (November 1976), 371–80; and Erwin Raisz, "The Rectangular Statistical Cartogram," *Geographical Review* 24, no. 2 (April 1934), 292–96.

6. If the mapping program does not allow the user to shift the positions of the area polygons, the map author should reopen the map file in a drawing program that allows polygons to be moved.

7. For discussion of the pitfalls of basing pattern analysis and correlation on a single level of aggregation, see Mark Monmonier, *How to Lie with Maps* (Chicago: University of Chicago Press, 1991), chap. 9.

8. For discussion of maps of residuals from linear regression, see Peter J. Taylor, *Quantitative Methods in Geography: An Introduction to Spatial Analysis* (Boston: Houghton Mifflin, 1977), 203–6; and John W. Tukey, *Exploratory Data Analysis* (Reading, Mass.: Addison-Wesley, 1977), 151–54.

9. For discussion of trend-surface analysis and spatial smoothing, see John C. Davis, *Statistics and Data Analysis in Geology*, 2d ed. (New York: John Wiley and Sons, 1986), 405–47.

10. See Richard Dunn, "Framed Rectangle Charts or Statistical Maps with Shading: An Experiment in Graphical Perception," *American Statistician* 42, no. 2 (May 1988), 123–29.

CHAPTER 7

1. Long chose these top five migration streams from among 2,256 interstate migration flows tabulated by the Bureau of the Census for the forty-eight conterminous United States. See Figure 3.1 in Larry Long, *Migration and Residential Mobility in the United States* (New York: Russell Sage Foundation, 1988), 64–65.

2. Robert Cervero, *Suburban Gridlock* (New Brunswick, N.J.: Center for Urban Policy Research, Rutgers University, 1986), 186–92.

3. Mark Monmonier and George A. Schnell, "Interstate Migration of Physicians in the U.S.: The Case of 1955–59 Graduates," *Professional Geographer* 28, no. 1 (February 1976), 29–34.

4. See U.S. Bureau of the Census, "Ratio of Workers Working in County to Workers Residing in County, in the United States: 1970," Map series GE-50, no. 63. A similar display map is not available for the 1980 census. The federal budget cuts of the early 1980s struck relatively harshly at the Census Bureau, and its excellent program of display maps was one casualty.

5. For a discussion of the "gravity model" and other interaction models used to adjust for distance, transport cost, and other factors that facilitate or retard flow among places, see Ronald L. Mitchelson and James O. Wheeler, "Analysis of Aggregate Flows," in Susan Hanson, ed., *The Geography of Urban Transportation* (New York: Guilford Press, 1986), 119–53; and Eric Sheppard, "The Distance Decay Gravity Model Debate," in Gary Gaile and Cort Willmot, eds., *Spatial Statistics and Models* (Dordrecht: D. Reidel, 1984), 367–88. For an example of a single-destination map adjusted for distance with an interaction model, see Mark Monmonier and Anthony V. Williams, "Abortion and Spatial Interaction: Temporary Migration to New York," *Proceedings of the Association of American Geographers* 5 (1973), 177–80.

6. In a paper examining the variety of graphic methods for representing spatial-temporal data with maps and graphs, I used the term "chess map" to relate these single-time cartographic snapshots to the pictorial diagrams that describe the board at discrete stages in a game of

chess. In the same article I invoked the term "dance map" to note the similarity between choreography diagrams used to teach ballroom dancing and maps with arrows or flow lines used to describe stepwise flows, military invasions, and movements along routes. See Mark Monmonier, "Strategies for the Visualization of Geographic Time-Series Data," *Cartographica* 27, no. 1 (Spring 1990), 30–45.

7. Kaniss, Phyllis, *Making Local News* (Chicago: University of Chicago Press, 1991), 35–42.

8. For an example, see U.S. Bureau of the Census, "1970 Population as a Percent of Maximum Population by Counties of the United States," Map series GE-50, no. 43.

9. For examples of some simple models for describing temporal change, see Colin Newell, *Methods and Models in Demography* (New York: Guilford Press, 1988), 180–89; and John Saunders, *Basic Demographic Measures: A Practical Guide for Users* (Lanham, Md.: University Press of America, 1988), 64–68.

10. Multidimensional scaling, an approach to creating a relative-space map reflecting interaction or rate structures for multiple origins and destinations, is beyond the scope of this book. For further information, see Jean-Claude Muller, "Non-Euclidean Geographic Spaces: Mapping Functional Distances," *Geographical Analysis* 14, no. 3 (July 1982), 189–203.

11. See, for example, the distance cartogram comparing parcel-post rates for two-pound and ten-pound parcels mailed from Syracuse in Mark Monmonier, *How to Lie with Maps* (Chicago: University of Chicago Press, 1991), 17. The rate structure for a ten-pound parcel yields a configuration of points that reflects shipping distance. In contrast, the two-pound rates somewhat reflect the flat rate structure for a first-class letter, which places all domestic destinations on the circumference of a single circle.

12. *Michelin Guide to the Battlefields of the World War,* vol. 1, *The First Battle of the Marne, 1914* (Milltown, N.J.: Michelin, 1919).

13. Keith Robbins, *The First World War* (Oxford: Oxford University Press, 1984), 35.

14. Sherry H. Olson, *Baltimore: The Building of an American City* (Baltimore: The Johns Hopkins University Press, 1980), x–xii.

CHAPTER 8

1. Ernest W. Burgess, "The Growth of the City: An Introduction to a Research Project," in Robert E. Park, Ernest W. Burgess, and Roderick D. McKenzie, *The City* (Chicago: University of Chicago Press, 1925), 47–62.

2. Homer Hoyt, *The Structure and Growth of Residential Neighborhoods in American Cities* (Washington, D.C.: Federal Housing Administration, 1939).

3. For the development of the multiple-nuclei model and an evaluation of the Burgess and Hoyt models, see Chauncy D. Harris and Edward L. Ullman, "The Nature of Cities," *Annals of the American Academy of Political and Social Science* 242 (1945), 7–17. For later graphic models addressing the influence of beltways and other transport corridors, see John S. Adams, "Residential Structures of Midwestern Cities," *Annals of the Association of American Geographers* 60 (1970), 37–62; and Peter O. Muller, "Transportation and Urban Form: Stages in the Spatial Evolution of the American Metropolis," in Susan Hanson, ed., *The Geography of Urban Transportation* (New York: Guilford Press, 1986), 24–48.

4. Mark Monmonier, "Railroad Abandonment in Delmarva: The Effect of Orientation on the Probability of Link Severance in a Transport Network," *Proceedings of the Pennsylvania Academy of Science* 44 (1970), 27–31.

5. My work on journalistic cartography required looking carefully at newspaper publishing as a business. I wrote four short papers on the geographic patterns of news publishing, one of which examined overlapping circulation zones. See Mark Monmonier, "Newspaper Circulation Areas in Central New York: A Geographic Refinement of the Umbrella Hypothesis," *Journal of the Pennsylvania Academy of Science* 64 (1990), 46–51.

6. James N. Rosse, *The Evolution of One-Newspaper Cities*, Stanford Studies in Industry Economics no. 95 (Stanford, Cal.: Stanford University Department of Economics, 1979), 5–9, 23–25.

7. For further discussion of the technology, underlying principles, and applications of geographic information systems, see David. J. Maguire, Michael F. Goodchild, and David W. Rhind, eds., *Geographical Information Systems: Principles and Applications* (London: Longman Group, 1991).

8. Barbara A. Anderson, *Internal Migration during Modernization in Late Nineteenth-Century Russia* (Princeton, N.J.: Princeton University Press, 1980), 127–33.

9. Hoyt, *Structure and Growth of Residential Neighborhoods*, 47.

10. See Ian R. Bartky and Elizabeth Harrison, "Standard and Daylight-Saving Time," *Scientific American* 240, no. 5 (May 1979), 46–53; and Ian R. Bartky, "The Adoption of Standard Time," *Technology and Culture* 30 (1989), 25–56.

11. For an example of four scatterplots with very different configurations, all reflecting a correlation coefficient of 0.5 for the same number of points, see F. J. Anscombe, "Graphs in Statistical Analysis," *American Statistician* 27 (1973), 17–21. The standard product-moment correlation coefficient evaluates a bivariate relationship according to only one statistical model, least-squares linear regression. A scatterplot allows the analyst to see whether another model might be more appropriate.

12. The term "geographic correlation" recognizes that a correlation based on spatial data has a geographic component that must be visualized and a statistical component that can be measured numerically. Geographic correlation is a narrower concept than the broader "geographic structure," which includes patterns and trends for a single attribute. For a perceptive essay on the importance of studying geographic structure, see Peter Gould, "Expose Yourself to Geographic Research," in John Eyles, ed., *Research in Human Geography: Introductions and Investigations* (Oxford: Basil Blackwell, 1988), 11–27.

13. For further discussion of map comparison based on dispersion breaks computed using the standard deviation, see R. W. Armstrong, "Standardized Class Intervals and Rate Computation in Statistical Maps of Mortality," *Annals of the Association of American Geographers* 67 (1977), 429–36.

14. Subject-testing experiments with college students suggest that similarity not in pattern but in the *amount* of area covered by darkest symbol might be the single most important influence on judgments of visual similarity by untrained map readers. See Robert Lloyd and Theodore Steinke, "Visual and Statistical Comparison of Choropleth Maps," *Annals of the Association of American Geographers* 67 (1977), 429–36.

15. See, for example, Richard Dunn, "A Dynamic Approach to Two-Variable Color Mapping," *American Statistician* 43 (1989), 245–52; and Judy M. Olson, "Spectrally Encoded Two-Variable Maps," *Annals of the Association of American Geographers* 71 (1981), 259–76.

16. For an introduction to exploratory data analysis, see William S. Cleveland, *Elements of Graphing Data* (Monterey, Calif.: Wadsworth, 1985); and John W. Tukey, *Exploratory Data Analysis* (Reading, Mass.: Addison-Wesley, 1977).

17. Least-squares regression was used to fit both the simple regression line in Figure 8.19 and the fourth-order polynomial in Figure 8.21. Coefficients of determination indicate that the polynomial fits the data much better ($r^2 = .376$) than the straight line ($r^2 = .144$). For discussion of the *lowess* technique, a more flexible line-fitting approach to data smoothing, see Cleveland, *Elements of Graphing Data*, 167–78.

18. For discussion of the effects of scale and aggregation on the cartographic analysis of correlation, see Mark Monmonier, *How to Lie with Maps* (Chicago: University of Chicago Press, 1991), 123–46.

19. See, for example, the essay "The Medium is the Message," in Marshall McLuhan, *Understanding Media: The Extensions of Man* (New York: New American Library, 1964), 23–39.

20. See Mark Monmonier, "Map-Text Coordination in Geographic Writing," *Professional Geographer* 33 (1981), 406–12.

21. Williams's writing guide is markedly more logical and useful than any other I have read, including the classic *Elements of Style* by

Strunk and White, and I recommend it highly. Although Williams says nothing about maps and diagrams per se, many of his concepts are as applicable to graphic discourse as they are to written. See Joseph M. Williams, *Style: Toward Clarity and Grace* (Chicago: University of Chicago Press, 1990), esp. chapter 3, "Cohesion."

APPENDIX A

1. All but a few of the maps and graphs in this book (except facsimiles) were drawn by the author using a Macintosh II computer and MacDraw II, a straightforward general-purpose drawing application.

APPENDIX B

1. I know of no official, frequently updated list of cartographic laboratories. Survey results and news items often appear in *Cartographic Perspectives*, the bulletin of the North American Cartographic Information Society. See, for example, Roy Doyon and Anne Gibson, "Academic Cartography Labs in the U.S. and Canada: A Survey," *Cartographic Perspectives* no. 5 (Spring 1990), 21–29. NACIS has no permanent address; a map librarian or interlibrary-loan officer might be helpful in obtaining this publication. I would also suggest consulting the annual directory of college and university departments published by the Association of American Geographers. This directory might be useful in locating a nearby graduate geography department with a cartographic laboratory, or at least with someone willing to make recommendations. Authors in other countries are advised to consult the person responsible for cartography at a nearby university geography department.

2. A significant source of the ideas for this section was William G. Loy, "Sample Cartography Lab Statement," *Cartographic Perspectives* no. 8 (Winter 1990–91), 12–13.

Sources of Illustrations

1.1 Alfred T. Mahan, *Sea Power and Its Relationship to the War of 1812*, vol. 2 (Boston: Little, Brown and Company, 1905), illustration facing p. 4.

1.2 Mahan, *Sea Power and Its Relationship to the War of 1812*, vol. 2, illustration facing p. 376.

1.3 Robert E. Park and Herbert A. Miller, *Old World Traits Transplanted* (New York: Harper and Brothers, 1921; Chicago: Society for Social Research, University of Chicago, 1925), 141.

1.4 Park and Miller, *Old World Traits Transplanted*, 201.

2.1 (upper) U.S. Geological Survey, 1979, Concord, Mass. 1:25,000-scale, 7.5-minute quadrangle map; (middle) U.S. Geological Survey, 1985, Boston, Mass., Rhode Island, Connecticut 1:100,000-scale planimetric map; (lower) U.S. Geological Survey, 1970, Boston, Mass., New Hampshire, Connecticut, Rhode Island, Maine 1:250,000-scale topographic map.

3.1 Derived from Jacques Bertin, *Semiology of Graphics: Diagrams, Networks, Maps*, trans. William J. Berg (Madison: University of Wisconsin Press, 1983), 43.

3.11 Compiled from maps in Harold M. Mayer and Richard C. Wade, *Chicago: Growth of a Metropolis* (Chicago: University of Chicago Press, 1969), 69, 108.

3.12 Redesigned from a page-size map titled "Circuity of Approach from South to Long Island via Selkirk," in United States Railway Association, *Preliminary System Plan*, vol. 2, *Restructuring Railroads in the Northeast and Midwest Region Pursuant to the Regional Rail Reorganization Act of 1973* (Washington, D.C., 26 February 1975), p. 364, fig. 4.

3.13 Redesigned from a figure in Wilbur Zelinsky, "Nationalism in the American Place-Name Cover," *Names* 31, no. 1 (March 1983), 23.

3.15 Derived from a map in James W. Cerny, "Joyce's Mental Map," *James Joyce Quarterly* 9, no. 2 (Winter 1971), 218–24.

4.11 Hans Kurath and others, *Linguistic Atlas of New England*, vol. 1 (Providence, R.I.: Brown University, 1939–43; copyright American Council of Learned Societies), map 11.

5.1 *County Atlas of Rensselear County, New York* (New York: F. W. Beers, 1876), 37.

5.3 U.S. Geological Survey, 1898, Syracuse, New York 1:63,360-scale, 15-minute quadrangle map.

5.4 U.S. Bureau of the Census, *Census of Agriculture, 1969,* vol. 5, *Special Reports,* pt. 15, *Graphic Summary* (Washington, D.C.: Government Printing Office, 1973), 34.

5.5 U.S. Department of State, *Geographic Notes,* no. 13 (1 March 1991), 37.

5.6 Collection of Civil War newspaper maps, Geography and Map Division, U.S. Library of Congress.

5.7 Compiled using Kokudo Chirin, *National Atlas of Japan* (Tokyo: Japan Map Center, 1977) and a clip-art map of Japan from MAP Art (Lambertville, N.J.: MicroMaps Software, 1990).

5.8 Photo 300-81 (August 1926 Airmap of City of Syracuse, N.Y., and Environs), Airmap Corp. of America, Brooklyn, N.Y.

7.1 Redesigned from a series of four maps presented at a larger scale on a different projection, in Larry Long, *Migration and Residential Mobility in the United States* (New York: Russell Sage Foundation, 1988), 64–65.

7.2 Robert Cervero, *Suburban Gridlock* (New Brunswick, N.J.: Center for Urban Policy Research, Rutgers University, 1986), 188. Copyright 1986, Rutgers—The State University of New Jersey, Center for Urban Policy Research. Reprinted with permission.

7.3 Redesigned from maps 1.1 and 1.2 in Phyllis Kaniss, *Making Local News* (Chicago: University of Chicago Press, 1991), 38.

7.5 Compiled from data in N. W. *Ayer and Son's American Newspaper Annual* (1900), and *Editor and Publisher Yearbook* (1990).

7.6 Compiled from airline advertising in mid-January 1991 and data in Jean Elizabeth Domey, "The Effect of Airline Deregulation on Air Service to Syracuse, New York" (Master's thesis, Syracuse University, 1979), 89.

7.7 *Michelin Guide to the Battlefields of the World War,* vol. 1, *The First Battle of the Marne, 1914* (Milltown, N.J.: Michelin, 1919), 17. Reprinted with permission.

7.8 U.S. Department of the Army, *American Military History: 1607–1953* (ROTC manual 145–20, July 1956), 396.

7.9 Keith Robbins, *The First World War* (Oxford: Oxford University Press, 1984), 35. Reprinted with permission.

7.10 Sections 6 and 8 of an eleven-map cartographic narrative in Sherry H. Olson, *Baltimore: The Building of an American City* (Baltimore: The Johns Hopkins University Press, 1980), x–xi. Reprinted with permission.

8.1 Robert E. Park, Ernest W. Burgess, and Roderick D. McKenzie, *The City* (Chicago: University of Chicago Press, 1925), 51. Reprinted with permission.

8.2 Park, Burgess, and McKenzie, *The City*, 55.

8.3 Homer Hoyt, *The Structure and Growth of Residential Neighborhoods in American Cities* (Washington, D.C.: Federal Housing Administration, 1939), 14.

8.4 Extracted and adapted from a diagram presenting sector models for 30 selected cities in Hoyt, *Structure and Growth of Residential Neighborhoods*, 77.

8.5 Extracted and adapted from a diagram presenting sector models for six selected cities in Hoyt, *Structure and Growth of Residential Neighborhoods*, 115.

8.6 Revised and updated from Mark Monmonier, "Railroad Abandonment in Delmarva: The Effect of Orientation on the Probability of Link Severance in a Transport Network" *Proceedings of the Pennsylvania Academy of Science* 44 (1970), 27–31. Originally compiled from various corporate and published sources.

8.9 Central portion of Figure 2 in Mark Monmonier, "Newspaper Circulation Areas in Central New York: A Geographic Refinement of the Umbrella Hypothesis," *Journal of the Pennsylvania Academy of Science* 64 (1990), 46–51. Reproduced with permission.

8.11 Barbara A. Anderson, *Internal Migration during Modernization in Late Nineteenth-Century Russia* (Princeton, N.J.: Princeton University Press, 1980), 128. Copyright 1980 by Princeton University Press. Reproduced with permission.

8.12 Hoyt, *Structure and Growth of Residential Neighborhoods*, 47.

8.14 Compiled and redesigned from "Superposed Time Zones" map in Ian R. Bartky and Elizabeth Harrison, "Standard and Daylight-Saving Time," *Scientific American* 240, no. 5 (May 1979), 47.

Index

Abbreviations, use as map symbols, 114–15

Additive overlays. *See* Cartographic overlays

Aerial photography: compilation from, 151–52; distortion of distance on, 151; as source of cartographic information, 132–34

Aesthetics: map labels, 108–9; map symbols, 75

Albers equal-area projection, 44–45

Alphabetic labels, as map symbols, 114–16

American Geographical Society: *Index to Maps in Books and Periodicals*, 137; map collection: 135–36; *Research Catalog*, 137

American Library Association, 135

Anchor stimuli, for graduated-point symbols, 63, 163

Anderson, Barbara, 221–22

Angles, distortion of, 32

Animated maps, 189

Anthropology, use of maps in, 14, 17, 29, 103, 245

Archaeology, use of maps in, 29, 77–78, 126

Area, distortion of, 33–35, 52

Area cartograms, 180–81, 185

Area symbols: aesthetics and, 75; for choropleth maps, 168; defined, 61; distinct from area features, 70; variety of, 70–76

Areal aggregation, effect of, 181–82, 240

Arrows, as cartographic symbols, 190–91, 202–3

Art history, use of maps in, 122, 226–27, 245

Articulacy, 9

ATLAS MapMaker (cartographic application software), 132

Atlas of Reproducible Pages, 143

Atlas of Switzerland, 128

Atlases: as sources, 128–30; thematic, 128–29; world, 128

Attribute space, 58, 228

Azimuthal projections, 42, 46–48, 52

Background information, 57, 68

Balloons, containing descriptive text, 205, 243

Bar scale. *See* Graphic scale

Bartky, Ian, 226–27

Battle fronts, cartographic representation of, 200–203

Bertin, Jacques, visual variables of, 57–58, 88, 106–7

Bibliographic Guide to Maps and Atlases, 136

Bibliographies. *See* Cartobibliographies

Bibliography of Cartography, 138

Bibliothèque Nationale (Paris), 156

Biographies, use of maps in, 204–5

Black, as a map symbol, 76, 80